NAVY SEALS BUG IN GUIDE

How to Transform Your Home into the Safest Place on Earth

JAMES LANDERS

© **Copyright 2024 by JAMES LANDERS - All rights reserved.**

All rights reserved. No part of this book may be reproduced in any form without permission in writing from the author. Reviewers may quote brief passages in reviews.

While all attempts have been made to verify the information provided in this publication, neither the author nor the publisher assumes any responsibility for errors, omissions, or contrary interpretation of the subject matter herein.

The views expressed in this publication are those of the author alone and should not be taken as expert instruction or commands. The reader is responsible for his or her own actions, as well as his or her own interpretation of the material found within this publication.

Adherence to all applicable laws and regulations, including international, federal, state, and local governing professional licensing, business practices, advertising, and all other aspects of doing business in the US, Canada or any other jurisdiction is the sole responsibility of the reader and consumer.

Neither the author nor the publisher assumes any responsibility or liability whatsoever on behalf of the consumer or reader of this material. Any perceived slight of any individual or organization is purely unintentional.

CONTENTS

INTRODUCTION 1
   The Value Of Bugging In: Why Remaining At Home Is The Safest Course Of Action 1
   Teachings From A Navy SEAL: Practical Survival Strategies 2

CHAPTER 1: Getting Your House Ready For Any Emergency 5
   Recognizing Possible Dangers 5
   Performing an Evaluation of Home Safety 9
   Formulating a Tailored Bug-In Strategy 12

CHAPTER 2: Food Storage For The Future 16
   Vital Foods for Survival 16
   Creating and Switching Up Your Food Stockpiles 20
   Advice on Appropriate Shelf Life and Storage 23

CHAPTER 3: Ensuring Access To Water Security 26
   Evaluating the Water Needs of Your Home 26
   Solutions for Safe Water Storage 28
   Methods of Filtration and Purification 31

CHAPTER 4: Creating Your Own Power 36
   Energy Requirements in an Emergency 36
   Off-Grid Power Options: Generators, Wind, and Solar 39
   Safety and Fuel Storage Issues 43

CHAPTER 5: Techniques For Home Défense 46
   Defending Your House: Windows, Doors, and the Exterior 46
   Tools and Techniques for Defense 49

CHAPTER 6: Interactions In Times Of Crisis 55
   Putting in Place Dependable Communication Mechanisms 55
   Digital security, radios, and emergency broadcasts 58
   Keeping Updated Without Internet Access 61

CHAPTER 7: Health Care Readiness And Basic First Aid 65
   Keeping Essential Medicines in Stock 65
   Putting Together a Complete First Aid Kit 68
   Medical Knowledge That Everyone Should Have 71

CHAPTER 8: Sustaining Hygiene And Sanitation 76
   Solutions for Waste Management for Extended Bug-Ins 76
   Maintaining Personal Hygiene: Keeping Clean and Well 79

    Avoiding Illness in Tight Spaces 82

CHAPTER 9: Crucial Skills For Survival 86

    Starting a Fire and Preserving Heat 86
    Basic Methods of Self-Defense 89
    Finding Your Way and Making Help Signals 93

CHAPTER 10: Psychological Resilience 97

    Dealing with Stress and Isolation 97
    Developing Mental Fortitude 100
    Keeping the Spirit Up in Extended Crises 103

CHAPTER 11: Monitoring And Security Systems 107

    Setting Up and Keeping an Eye on Home Security Cameras 107
    Alarms, Motion Sensors, and Integrations with Smart Homes 111
    Establishing a Network for Neighbourhood Watch 114

CHAPTER 12: Sharing Of Resources And Communities 119

    Working Together for Mutual Aid with Neighbours 119
    Establishing a Community of Support 122
    Trading and Safe Resource Sharing 126

CHAPTER 13: Superior House Defense 130

    Strengthening the Structural Integrity of Your House 130
    Bulletproofing and Blast Protection Options 133
    Establishing Covered Safe Areas 136

CHAPTER 14: Adjusting To Shifting Circumstances 139

    Adapting Your Strategy to Various Situations 139
    Reacting to Novel Dangers and Obstacles 142
    Acquiring Knowledge from Actual Bug-In Incidents 145

CONCLUSION 148

INTRODUCTION

The Value Of Bugging In: Why Remaining At Home Is The Safest Course Of Action

During a crisis, the natural inclination is to flee, to distance oneself as much as possible from the imminent threat. However, the act of escaping to an unfamiliar location can actually expose you to additional hazards and difficulties for which you may not have adequate preparation. Therefore, in the majority of situations, choosing to stay in your own home, often known as bugging-in, is not only a feasible choice but also frequently the most secure and efficient one.

Your residence serves as a fortress of protection and security. It is the place where you may easily get and use the resources that you have collected over many years, including your food, water, tools, and medical supplies. You possess superior knowledge of your residence compared to any other location on the planet. By comprehending its advantages and disadvantages, you can utilize that understanding to establish a customized and safe setting that caters to your particular requirements.

Reasons Why It Could Be Dangerous to Leave

Take into account the sacrifices you make while deciding to evacuate: the luxury of a cosy sleeping arrangement, the protection of secured entrances, and the sense of familiarity with your environment. Upon exiting, you are left to navigate an environment that may pose a threat or even be life-threatening. Unforeseen hazards like as severe weather, hazardous landscapes, or contacts with desperate individuals seeking survival may be encountered.

Without a clearly defined and carefully thought-out objective, the act of evacuating can rapidly devolve into a highly distressing and chaotic situation. One may encounter situations where there is a scarcity of food or water, difficulty in locating suitable shelter, or even endangering oneself and loved ones.

The Benefits of Staying at Home

On the other hand, choosing to bug-in enables you to take full advantage of all the resources and advantages that your home provides. By remaining in one place, you may strengthen your living area, transforming it into a stronghold of safety and protection. You possess the necessary time and resources to strategically devise, organize, and safeguard your residence from prospective hazards. Your home provides you with the highest level of control over your circumstances, whether it be during natural disasters, civil upheaval, or prolonged power outages.

By staying at home, you have the opportunity to establish and sustain relationships with the people in your neighborhood. During periods of turmoil, possessing a reliable group of trusted neighbors might be crucial. By collaborating, you can pool resources, offer reciprocal assistance, and enhance your combined safety.

Realistic Thinking Over Fantasy

Although the concept of bugging out may appear thrilling, involving a self-sufficient lifestyle and dependence on survival abilities, the actuality is much distinct. Thriving in the wilderness necessitates profound expertise, exceptional physical stamina, and a substantial amount of serendipity. Experienced survivalists recognize the significance of establishing a stable and safe headquarters. The stability that your house provides, with its abundant resources and familiar surroundings, is invaluable.

Choosing to bug-in is not just about ensuring survival, but also about maintaining a sense of normality and comfort in a chaotic world. It concerns the ability to safeguard your family, preserve your well-being, and provide housing stability among widespread disorders.

The Astute Decision

In most circumstances, choosing to bug-in is ultimately the intelligent decision for the majority of individuals. It involves employing critical thinking, engaging in proactive planning, and optimizing the utilization of available resources. In this book, I will provide you with a comprehensive guide on the necessary procedures to transform your home into the most secure location on the planet. Upon completion, you will possess the expertise and self-assurance to confront any emergency with composure and conviction, fully aware that you have made the optimal decision for yourself and your loved ones.

Teachings From A Navy SEAL: Practical Survival Strategies

Surviving in the presence of hardship is not solely dependent on possessing appropriate equipment or superior provisions—it hinges on possessing the appropriate mentality, adequate training, and the capacity to adjust to swiftly evolving circumstances. Being a Navy SEAL, I have received extensive training to effectively navigate and manage very perilous and challenging conditions seen in various harsh regions across the globe. The knowledge I have acquired is not only theoretical but has been put to the test and validated in real-life scenarios where the consequences were a matter of life or death.

The Mindset of a Survivor

An essential lesson I can impart is that survival originates in the mind. During a crisis, it is common to be overwhelmed by anxiety, uncertainty, and doubt. However, during those instances, it is imperative to maintain composure, concentration, and resolve. The mindset of a survivor is recognizing that fear is detrimental and that clear, logical thinking is the most potent tool at one's disposal.

As a member of the Navy Seals, I received training to anticipate unforeseen events, to be ready for the most unfavorable circumstances, and to consistently have a well-thought-out strategy. Furthermore, I received instruction on being flexible, thinking swiftly, and making prompt and assured decisions, especially in situations where complete knowledge was lacking. These skills are universally attainable and crucial for safeguarding your residence and loved ones.

The Value of Being Ready

Within the SEALs, we adhere to the adage: "The greater your exertion during training, the lesser your likelihood of injury during combat." Thorough preparation is crucial. Prior to being able to adequately address a crisis, it is imperative to have made the necessary preparations. This entails not just accumulating provisions but also honing the necessary abilities to employ them. It entails strategizing for all conceivable situations, so enabling you to be prepared and aware of the appropriate course of action in the event of any mishap.

Preparation entails not just having a strategy, but also having contingency plans and additional backups. It involves meticulously analysing every aspect, contemplating all potential outcomes, and ensuring that you possess the necessary resources and expertise to tackle any situation that arises.

Changing with Your surroundings

The uniqueness of each crisis necessitates the importance of your adaptability to ensure your survival. During our training in the SEALs, we were exposed to several habitats such as deserts, jungles, mountains, and urban settings. This was done intentionally as each environment poses distinct obstacles. This also applies to your residence. In the face of a natural disaster, power loss, or civil upheaval, it is crucial to modify your survival strategy according to the exact circumstances at hand.

This entails comprehending your surroundings, recognizing the advantages and limitations of your residence, and being prepared to adapt your strategies as necessary. Additionally, it entails being resourceful, devising innovative resolutions to challenges, and utilizing available resources to fulfil your requirements.

The Value of Teamwork

During our time in the SEALs, we always operated as part of a team and never embarked on a mission individually. Collaboration was crucial to our achievement, and it holds equal significance in a scenario where survival is at stake. Whether you are collaborating with your family, your neighbours, or a wider community, having a dependable team is crucial. An well synchronized team can distribute the effort, offer reciprocal assistance, and enhance the likelihood of enduring any emergency.

However, effective collaboration also necessitates confidence, effective communication, and a collective dedication to the shared objective of ensuring survival. Prior to hunkering down, consider strategies for establishing and fortifying your network of assistance. Collaboration, by means of resource sharing, task delegation, and mutual support, will greatly benefit you as a team.

Real-World Experience

The methods and ideas presented in this book are derived not only from theoretical study, but also from practical experience. I have encountered circumstances in which survival was not optional, but rather imperative. I have observed both effective and ineffective strategies. I have discovered that the most effective method for enduring is to be well-prepared, maintain composure, and continuously adjust to new circumstances.

Upon completing this book, you will possess the necessary information and abilities to transform your residence into an impregnable stronghold, safeguard your cherished individuals, and endure any challenges that may arise in the world. The knowledge gained from my experience as a Navy SEAL has been acquired via difficult challenges, and I am now imparting it to you so that you can be prepared for any situation.

CHAPTER 1
GETTING YOUR HOUSE READY FOR ANY EMERGENCY

Recognizing Possible Dangers

Efficient crisis preparedness commences with a comprehensive comprehension of the potential risks that you might encounter. Through the process of identifying and analysing these dangers, you may enhance your preparation efforts by focussing on addressing specific hazards. This section aims to provide you with a comprehensive understanding of different sorts of dangers, enabling you to evaluate their potential consequences on your house, and devise effective methods to minimise their impact.

1. Assessing Regional Risks

The geographical location you are in has a substantial impact on the specific types of dangers you may come into. Gaining insight into local hazards enables you to allocate your readiness endeavours according to the most probable circumstances.

- **Natural Disasters:** Evaluate the prevalent natural dangers in your region:
  - **Hurricanes/Typhoons:**
    - **Impact**: Flooding and structural damage can be caused by high winds, storm surges, and severe rains.
    - **Preparation**: Strengthen windows and doors, get an alternative power supply, and accumulate non-perishable food and water.
- **Earthquakes:**
  - **Impact**: Seismic activity can result in structural failures, conflagrations, and gas seepage.
  - **Preparation**: Ensure the stability of bulky furniture and appliances, strengthen the structural soundness of your residence, and prepare an emergency pack.

- **Tornadoes:**
    - **Impact:** Homes can be destroyed, and injuries can occur as a result of extremely strong winds.
    - **Preparation:** Construct a tornado shelter or a secure room and equip it with essential emergency provisions.
- **Floods:**
    - **Impact:** Water damage has the potential to devastate property and provide significant health risks.
    - **Preparation:** Enhance electrical systems, deploy sump pumps, and store crucial documents in waterproof containers.
- **Wildfires:**
    - **Impact:** Homes, particularly those located in wooded or brush-covered regions, might be endangered by fire and smoke.
    - **Preparation:** Establish a perimeter of protection around your residence, employ materials that are resistant to fire, and develop a detailed strategy for evacuating.
- **Severe Winter Weather:**
    - **Impact:** Severe weather conditions such as snow, ice, and extremely low temperatures have the potential to disrupt utilities and create dangerous situations.
    - **Preparation:** Ensure proper insulation in your home, maintain a stock of warm clothes, and have emergency heating options readily accessible.
- **Man-Made Disasters:** Take into account the potential hazards that result from human actions:
    - **Industrial Accidents:**
        - **Impact:** Chemical spills or explosions originating from adjacent companies or plants.
        - **Preparation:** Implement air filtration systems, establish an evacuation strategy, and maintain a supply of personal protective equipment.
- **Nuclear or Radiological Events:**
    - **Impact:** Exposure to radiation resulting from accidents or deliberate attacks.
    - **Preparation:** Ensure you have a well-defined strategy for either remaining in a secure location or relocating, and accumulate a enough supply of potassium iodide tablets.
- **Terrorism:**
    - **Impact:** Potential hazards encompassing explosions, active shooter incidents, or other forms of assaults.

- **Preparation:** Improve residential security, create communication protocols, and formulate tactics for implementing lockdown or evacuation procedures.

- **Economic Collapse:**
    - **Impact:** Service and supply chain disruptions caused by financial volatility.
    - **Preparation:** Accumulate crucial provisions, vary your sources of sustenance and hydration, and establish alternate methods of obtaining financial assets.

2. Evaluating Home Vulnerabilities

Gaining insight into the vulnerabilities of your residence in connection to these hazards enables you to implement specific enhancements to bolster security and durability.

- **Structural Integrity:** Examine your house for any flaws.
    - **Foundation:** Inspect for fissures or indications of instability. Make careful to properly maintain and strengthen the foundation if necessary.
    - **Roofing:** Conduct a thorough examination to identify any signs of damage or leaks. Ensure that your roof is equipped to endure extreme weather conditions.
    - **Walls and Windows:** Make sure that the walls are sturdy and that the windows are strengthened. If deemed required, proceed with the installation of impact-resistant windows or shutters.

- **Location-Specific Risks:** Evaluate potential hazards according to the geographical position of your residence:
    - **Flood Zones:** Assess whether your property is located in a region that is susceptible to flooding. Please prioritise the elevation of key systems and the implementation of drainage solutions.
    - **Proximity to Hazardous Facilities:** Assess the potential hazards posed by neighbouring industrial or chemical facilities. Make sure you have sufficient safeguards and strategies in place for protection and evacuation.
    - **Neighborhood Security:** Take into account the security of your local area. Implement security measures and foster connections with neighbours to bolster community safety.

3. Historical Data and Trends

An analysis of historical data offers valuable insights into the specific types of situations that are most probable to transpire in your vicinity, enabling you to proactively predict and address future challenges.

- **Local History:** Conduct a thorough investigation of previous calamities and disasters that have occurred in your local area. This includes:
    - **Natural Disasters:** The frequency and magnitude of previous natural occurrences, such as hurricanes, earthquakes, and floods.

- **Economic Events:** An analysis of past financial crises and their impact on local services and infrastructure.

- **Man-Made Incidents:** Documentation of industrial accidents, terrorist assaults, or other notable incidents.

- **Trends and Patterns:** Analyse patterns in the occurrence and intensity of these incidents. Search for recurring patterns such as:

  - **Increasing Frequency:** Are some categories of disasters experiencing a higher frequency of occurrence?

  - **Severity Changes:** Is the magnitude of these occurrences escalating with time?

4. Creating a Risk Profile

Utilising your evaluation of area hazards, vulnerabilities within your home, and past data, generate a risk profile to direct your preparedness endeavours.

- **High-Risk-Areas:** Determine the most critical risks that necessitate quick action. Formulate precise tactics to mitigate these hazards.

- **Moderate-Risk-Areas:** Once the high-priority areas have been addressed, proceed to tackle these threats. Develop a strategy to reduce the impact, but distribute resources appropriately.

- **Low-Risk Areas:** Formulate a strategy to address these hazards, although they may not necessitate as comprehensive preparation as more important concerns.

5. Regular Reassessment

Regular reassessment is crucial for sustaining effective readiness due to the evolving nature of threats and vulnerabilities.

- **Annual Reviews:** Perform a thorough assessment of the weaknesses and possible dangers to your residence on an annual basis.

- **Adapt to Changes:** Revise your readiness strategies and provisions in response to alterations in your risk assessment or fresh insights into impending hazards.

This section offers an in-depth explanation of how to identify possible dangers, evaluate weaknesses in one's house, and create a risk profile to improve readiness.

Performing an Evaluation of Home Safety

Conducting a comprehensive home safety evaluation is essential for identifying weaknesses and ensuring that your home is adequately prepared for any emergency. This procedure entails assessing the structural soundness of your residence, fortifying access points, and implementing essential enhancements to improve safety and durability. This section will provide you with a step-by-step instruction on how to conduct a thorough home safety evaluation.

1. Structural Integrity

Evaluating the structural soundness of your residence is crucial in guaranteeing its ability to endure a wide range of catastrophes, including both natural calamities and human-caused events.

- **Foundation:**
    - **Inspection:** Inspect the foundation for any signs of cracks, shifting, or settling. Minor fissures are frequently observed, however more substantial fractures may suggest significant problems.
    - **Reinforcement:** If you detect notable issues, it is advisable to strengthen the foundation by seeking assistance from professionals in order to avert additional harm.
- Roofing:
    - **Inspection:** Inspect for any absent or deteriorated shingles, evidence of water leakage, and indications of deterioration. Ensure that the roof is capable of withstanding severe weather conditions such as heavy snowfall or strong winds.
    - **Maintenance:** Periodically maintain gutters and downspouts to avoid water-related harm. If you reside in a high-risk area, it is advisable to upgrade to roofing materials that are resistant to impact.
- Walls And Windows:
    - **Inspection:** Inspect the walls for any signs of cracks or damage. Make sure that the outer walls are resistant to weather conditions and can endure strong winds or earthquake events.
    - **Reinforcement:** Install windows and doors that are resistant to impact and use weather stripping to prevent drafts and water from entering. Please contemplate the inclusion of security bars or shutters to enhance the level of protection.

2. Security and Entry Points

Securing access points is crucial for safeguarding your residence from unauthorised entrance and ensuring a safe means of evacuation, if necessary.

- **Doors:**
    - **Inspection:** Verify that all outside doors are constructed of sturdy materials and are correctly positioned. Verify that door frames are stable and devoid of any signs of harm.

- **Reinforcement:** Implement robust deadbolts, fortify door frames, and contemplate integrating security cameras or motion-activated lighting at entryways.

- **Windows:**
    - **Inspection:** Ensure that all windows are properly fastened and devoid of any fractures or impairments. Ensure that they function correctly while opening and closing.
    - **Reinforcement:** Please contemplate the inclusion of window security films, bars, or shutters. Utilise window locks and verify their functionality.

- **Garage and Other Access Points:**
    - **Inspection:** Ensure the security of your garage doors and any other points of entry, such as basement or attic access points.
    - **Reinforcement:** Implement robust locking mechanisms, fortify entry points, and employ motion sensors for surveillance in these designated zones.

3. Routes for Evacuation and Emergency Exits

It is crucial to have many emergencies exits and unobstructed evacuation pathways in order to ensure safety during a crisis.

- **Identification:**
    - **Main Exits:** Ensure that the primary egress points are easily reachable and clear of any obstacles. This encompasses both the front and back entrances.
    - **Secondary Exits:** Identify supplementary egress points, such as windows or side doors, that can be utilised in the event that the main exits are obstructed.

- **Accessibility:**
    - **Clear Pathways:** Ensure that all exit routes and pathways are free from any clutter or barriers. Frequently inspect for any obstacles and make any required modifications.
    - **Signage:** Employ conspicuous, discernible indicators to designate emergency egress points and evacuation pathways, particularly within expansive or multi-level residences.

4. Supplies and Equipment for Safety

Having appropriate safety equipment and supplies is crucial for properly handling crises.

- **Fire Extinguishers:**
    - **Placement:** Place fire extinguishers strategically in critical locations, such as the kitchen, garage, and in proximity to heating apparatus.

- **Maintenance:** Periodically inspect fire extinguishers to ensure they are in optimal operational state and replace them as necessary.

- **Smoke and Carbon Monoxide Detectors:**
  - **Installation:** Install smoke alarms in each bedroom, corridor, and communal space. Place carbon monoxide detectors in close proximity to sleeping areas.
  - **Testing:** Conduct regular monthly tests on detectors and replace batteries on a yearly basis. Verify the proper functionality of the detectors.

- **First Aid Kits:**
  - **Contents:** Ensure that your first aid kits are stocked with necessary provisions, like bandages, antiseptics, analgesics, and pharmaceuticals.
  - **Location:** Place first aid kits in readily accessible areas and ensure that all members of the family are aware of their placement.

5. Utilities and Systems

Ensuring the security and functionality of your home's utilities and systems can enhance your ability to properly handle crises.

- **Electrical System:**
  - **Inspection:** Ensure that you engage the services of a certified electrician to thoroughly examine your electrical system for both safety and operational efficiency. Search for potential dangers such as uncovered electrical cables or circuits that are carrying too much load.
  - **Backup Power:** It is advisable to install a backup power source, such as a generator, in order to ensure uninterrupted power supply during periods of power outages.

- **Plumbing:**
  - **Inspection:** Inspect your plumbing system for any signs of leaks, corrosion, or other potential problems. Verify that your water supply and drainage systems are operating correctly.
  - **Shut-Off Valves:** Acquaint yourself with the whereabouts of water shut-off valves and verify that they are easily reachable in the event of an emergency.

- **Heating and Cooling Systems:**
  - **Inspection:** Periodically maintain your heating and cooling systems to guarantee their optimal functionality.
  - **Maintenance:** To provide a secure and pleasant environment, it is necessary to replace filters, clean vents, and resolve any problems with HVAC systems.

6. Regular Updates and Drills

Maintaining a regular schedule for updating safety measures and performing drills is crucial to maintain the ongoing effectiveness of your readiness.

- **Review and Update:** Regularly reassess and revise your home safety evaluation in response to modifications in your living environment, emerging hazards, or insights gained from practice exercises.
- **Drills:** Engage in frequent exercises with your family to rehearse emergency protocols and guarantee that each individual is familiar with their assigned tasks and obligations.

Formulating a Tailored Bug-In Strategy

An intricately devised bug-in strategy guarantees that you and your family are adequately prepared to remain securely and comfortably in your residence during any emergency situation. This strategy should encompass all facets of your home's preparedness, ranging from procuring vital provisions to establishing effective communication tactics. This section will provide step-by-step instructions on how to create a customised bug-in plan that is specifically designed to meet your own requirements and situation.

1. Establish Your Goals and Objectives

Commence by unambiguously describing your aims and targets for your bug-in strategy. This will aid in directing your attention towards the most crucial elements and guarantee that every piece of your plan is in harmony with your overarching preparedness approach.

- **Safety and Security:** Give top priority to ensuring the safety and security of your house and family. This encompasses the tasks of ensuring the safety of access points, devising strategies for emergency exits, and preparing for any threats.
- **Comfort and Sustainability:** Ensure the maintenance of comfort and functionality in your home during a crisis. Strategy for regulating temperature and ensuring a secure living space.
- **Self-Sufficiency:** Strive for self-sufficiency by ensuring you have necessary provisions, producing electricity, and effectively managing water resources. This will assist in diminishing reliance on external resources.

2. Evaluate Your Available Resources

Assess the existing resources at your disposal and identify any deficiencies that must be resolved in your bug-in strategy.

- **Food and Water Supply:**
    - **Inventory:** Assess your existing food and water provisions. Evaluate the duration of their longevity and determine if they fulfil the requirements of your family.

- **Expansion:** Identify supplementary items to accumulate, such as non-perishable edibles, water, and specialised dietary provisions.

- **Power and Heating/Cooling:**
  - **Backup Power:** Assess your existing power sources and contingency plans, such as generators or solar panels.
  - **Heating/Cooling:** Evaluate your heating and cooling systems and determine any additional actions required to ensure a pleasant temperature.

- **Safety and Security Equipment:**
  - **Inventory:** Inspect the state and whereabouts of safety apparatus, such as fire extinguishers, smoke detectors, and first aid kits.
  - **Upgrades:** Determine any supplementary safety or security precautions required, such as fortified entrances, surveillance systems, or additional power sources.

3. Develop Emergency Protocols

Develop explicit emergency procedures to direct you and your family in the event of a disaster. This will facilitate the dissemination of information and promote a clear understanding of the necessary actions and safety measures.

- **Evacuation Plans:**
  - **Routes:** Create a comprehensive plan for multiple evacuation routes starting from your residence. Determine secure areas where you can seek refuge in the event of needing to evacuate.
  - **Family Roles:** Allocate roles and responsibilities to each member of the family, specifying the individuals responsible for carrying out particular chores in the event of an evacuation.

- **Communication Plans:**
  - **Contact Information:** Create a comprehensive roster of essential contacts, encompassing relatives, neighbours, and emergency service providers. Ensure universal access to this information.
  - **Communication Methods:** Develop protocols for maintaining communication during an emergency, such as utilising radios, satellite phones, or messaging applications.

- **Medical and First Aid:**
  - **Medical Needs:** Please provide information about any particular medical requirements or conditions that exist within your family. Make sure you own the essential prescriptions and medical supplies.
  - **First Aid Training:** Make sure that there is at least one individual in the home who has received training in first aid and CPR. If necessary, it is advisable to pursue more training.

4. Stockpile Essential Supplies

Establishing a comprehensive stockpile of necessary provisions is vital for achieving sustained independence in the event of a bug-in scenario.

- **Food Supplies:**
    - **Types of Food:** Accumulate non-perishable food items, such as canned products, dehydrated fruits, cereals, and sources of protein.
    - **Storage:** Keep food in a cold and dry location. Utilise hermetically sealed containers to maintain the freshness of commodities and prevent any form of contamination.

- **Water Supplies:**
    - **Quantity:** Strive to maintain a reserve of at least one gallon of water per individual every day for a duration of at least two weeks.
    - **Purification:** Ensure access to clean water by using water purification techniques, such as the use of filters or purification tablets.

- **Household Items:**
    - **Cleaning Supplies:** Acquire an ample supply of cleaning agents, disinfectants, and sanitation materials.
    - **Personal Hygiene:** Include personal hygiene products, such as cleansing agents, dental care products, and sanitary items.

5. Prepare for Power and Utilities

Make sure you have dependable power and utility solutions in position to uphold comfort and functionality during a crisis.

- **Backup Power:**
    - **Generators:** When utilising a generator, make sure you adequately maintain it and have a sufficient fuel supply.
    - **Alternative Sources:** Explore alternative energy sources, such as solar panels or battery-operated equipment.

- **Water Management:**
    - **Storage:** Store water in sizable containers to guarantee an ample supply.
    - **Conservation:** Adopt water conservation measures to prolong your water resources.

- **Heating and Cooling:**
    - **Maintenance:** Consistently perform maintenance on your heating and cooling systems to guarantee optimal functionality.
    - **Alternative Solutions:** Explore other methods of heating or cooling, such as utilising wood fires or employing portable fans.

6. Evaluate and Modify Your Plan

Consistently conducting tests and making revisions to your bug-in plan guarantees its ongoing effectiveness and relevance.

- **Drills:** Regularly engage in drills with your family to rehearse emergency procedures and guarantee that everyone is acquainted with their assigned tasks and obligations.
- **Review:** Regularly reassess and revise your plan in response to new facts, modifications in your home, or developing dangers.
- **Feedback:** Collect input from family members following exercises or actual events to pinpoint areas that might be enhanced.

CHAPTER 2
FOOD STORAGE FOR THE FUTURE

Vital Foods for Survival

Accumulating the appropriate food items is essential to guarantee that you and your family possess the essential nutrients and vitality to maintain good health over a prolonged emergency. This section provides a comprehensive list of the necessary food items to include in your long-term food stockpile. The emphasis is on selecting things that offer a combination of nutritional value, extended shelf life, and adaptability.

1. Proteins

Proteins play a crucial role in preserving muscle mass, supporting immunological function, and promoting overall health.

- **Canned Meats:**
  - **Options:** Include tinned chicken, beef, pig, and fish. These are a valuable protein source and may be utilised in a variety of ways.
  - **Storage:** Verify that the cans are intact and stored in a cool and dry location. Regularly monitor expiration dates.

- **Dry Beans and Lentils:**
  - **Varieties:** Acquire a sufficient supply of legumes, including black beans, kidney beans, chickpeas, and lentils. They provide a high content of protein and fibre.
  - **Preparation:** Beans and lentils can be prepared in big quantities and utilised in soups, stews, and salads.

- **Nut Butters:**
  - **Types:** Peanut butter and almond butter are rich in protein and contain beneficial lipids.
  - **Storage:** To prevent deterioration, it is advisable to store the items in containers that are airtight and to keep them in a location that is cool and devoid of light.

- **Powdered or Canned Eggs:**
  - **Uses:** Powdered or canned eggs are highly adaptable and offer a valuable source of protein for various baking and cooking purposes.
  - **Storage:** Store in a cold and dry location, making sure that the packing remains undamaged.

2. Carbohydrates

Carbohydrates are crucial for supplying necessary energy and regulating blood sugar levels.

- **Rice:**
  - **Types:** Choose from a selection of white, brown, or quick rice. It possesses a considerable duration of storage and serves as a fundamental component for several dishes.
  - **Storage:** In order to prolong the duration for which the things can be stored, it is advisable to place them in containers that are airtight or in Mylar bags along with oxygen absorbers.
- **Pasta:**
  - **Varieties:** Acquire a variety of pasta varieties, such as spaghetti, macaroni, and noodles. Pasta is conveniently stored and prepared.
  - **Storage:** Store in a cold and dry location and inspect for any indications of pests or deterioration.
- **Oats and Cereals:**
  - **Options:** Rolled oats and whole-grain cereals are highly nutritious, providing ample amounts of carbohydrates and fibre.
  - **Storage:** Ensure that you store items in containers that are tightly sealed to avoid the entry of moisture and pests.
- **Flour:**
  - **Types:** Flours with versatile applications, such as all-purpose flour, whole wheat flour, and speciality flours like gluten-free, are suitable for both baking and cooking purposes.
  - **Storage:** Preserve your items by storing them in containers that are airtight or sealed with a cover to prevent damage from bugs and moisture.

3. Fruits and Vegetables

Fruits and vegetables offer important vitamins, minerals, and fibre.

- **Canned Vegetables:**
  - **Options:** Incorporate a diverse selection of vegetables, such as corn, green beans, carrots and tomatoes.

- **Storage:** Verify the integrity and expiration dates of the cans. Periodically rotate inventory.
- **Freeze-Dried Fruits:**
  - **Types:** Freeze-dried apples, berries, bananas, and other fruits preserve the majority of their nutrients and flavour.
  - **Storage:** Place the items in hermetically sealed containers or Mylar bags along with oxygen absorbers.
- **Powdered Vegetables:**
  - **Uses:** Vegetable powders are capable of augmenting the flavour of soups, stews, and sauces.
  - **Storage:** Store in hermetically sealed containers in a cool and arid location.
- **Dehydrated Vegetables:**
  - **Options:** Dehydrated vegetables, such as potatoes, carrots, and bell peppers, has the qualities of being lightweight and convenient for storage.
  - **Storage:** Preserve the dryness of the contents by storing them in hermetically sealed containers.

4. Fats and Oils

Fats and oils play a crucial role in providing energy and promoting general well-being.

- **Cooking Oils:**
  - **Types:** Incorporate olive oil, coconut oil, and vegetable oil. These oils serve distinct purposes in culinary and food processing.
  - **Storage:** Keep in a cool and dark location. Periodically rotate supplies to maintain optimal freshness.
- **Shelf-Stable Margarine or Butter:**
  - **Uses:** Margarine and butter are both suitable for use in baking and cooking.
  - **Storage:** Store in hermetically sealed containers and adhere to the storage guidelines provided by the manufacturer.

5. Dairy and Dairy Alternatives

Dairy products offer a rich source of calcium and other vital minerals.

- **Powdered Milk:**
  - **Uses:** Powdered milk is a very adaptable substitute for fresh milk that can be employed in various culinary applications, such as baking and frying.
  - **Storage:** Preserve in hermetically sealed containers in a cool and arid location.

- **Shelf-Stable Milk:**
  - **Types:** UHT (ultra-high temperature) milk can be stored at room temperature until it is opened, without the need for refrigeration.
  - **Storage:** Store in a cool and dry location. Regularly monitor expiration dates.
- **Dairy Alternatives:**
  - **Options:** For individuals who are lactose intolerant or have dietary limitations, it is advisable to incorporate powdered or canned substitutes such as soy or almond milk.
  - **Storage:** Adhere to the storage guidelines provided by the manufacturer.

6. Emergency Essentials

These goods are essential for preserving health and handling emergencies.

- **Salt and Spices:**
  - **Uses:** Salt and spices augment taste and are vital for food preservation.
  - **Storage:** Preserve the effectiveness and avoid moisture by storing in containers that are airtight.
- **Honey:**
  - **Uses:** Honey is a natural sweetener that has a prolonged storage period and can also serve as a natural cure for specific conditions.
  - **Storage:** Store in hermetically sealed containers and keep in a cool and arid location.
- **Vitamins and Supplements:**
  - **Types:** Ensure you have an ample supply of vital vitamins and supplements to address any deficiencies in your diet.
  - **Storage:** Keep in their original packing in a cool and dry location.

7. Beverages

Drinks play a crucial role in maintaining hydration and providing a range of options.

- **Instant Coffee or Tea:**
  - **Types:** For a diverse range of choices, consider including instant coffee, tea bags, or powdered drink mixes.
  - **Storage:** Store in hermetically sealed containers to avoid humidity and maintain the quality of the product.

- **Electrolyte Powders:**
  - **Uses:** Electrolyte powders aid in preserving hydration and restoring vital nutrients.
  - **Storage:** Preserve in hermetically sealed containers in a cool and arid location.

Creating and Switching Up Your Food Stockpiles

Establishing a complete stockpile of food is crucial for ensuring prolonged life in the event of a crisis. An equally crucial aspect is the effective management and periodic rotation of your food supplies to guarantee their freshness and usability. This section will provide you with a comprehensive guidance on how to efficiently construct and sustain your food stockpile.

1. Developing Your Food Stockpile

Establishing a comprehensive food stockpile requires strategic planning and meticulous selection of supplies.

- **Determine Your Stockpile Goals:**
  - **Duration:** Determine the desired duration for which you intend to maintain your food stockpiles. It is advisable to have a minimum of a 3-month stockpile, but you may need to consider longer lengths depending on your specific situation.
  - **Family Size:** Determine the quantity of food required by considering the number of individuals in your family and their daily caloric and nutritional requirements.

- **Create a Food Inventory List:**
  - **Categories:** Categorise your inventory list according to food groups, including proteins, carbohydrates, vegetables, fruits, fats, dairy, and emergency supplies.
  - **Quantities:** Provide the specific quantities required for each item, taking into account both individual portions and larger quantities suitable for long-term preservation.

- **Prioritize Food Types:**
  - **Staples First:** Commence by accumulating essential food items that possess a prolonged shelf life and serve as the foundation of your meals, such as rice, pasta, and canned goods.
  - **Supplementary Items:** Incrementally incorporate additional goods such as spices, condiments, and comforting foods to augment diversity and boost morale.

- **Source Your Supplies:**
  - **Bulk Purchases:** Acquire non-perishable goods in large quantities to economise and guarantee sufficient stock. Seek out promotions and markdowns to optimise the value obtained.

- **Long-Term Storage Options:** It is advisable to invest in durable food items for long-term storage, such as freeze-dried meals, MREs (Meals Ready-to-Eat), and Mylar bags equipped with oxygen absorbers to ensure the best possible preservation.

2. How to Store Your Groceries

Effective storage is essential for preserving the quality and ensuring the safety of your food supplies.

- **Storage Conditions:**
  - **Temperature:** Store food in a location that is cold, dry, and maintains a steady temperature. Steer clear of regions that are susceptible to temperature variations or intense heat.
  - **Humidity:** Store food in a low-humidity setting to inhibit the growth of mould and prevent it from spoiling.
- **Storage Containers:**
  - **Airtight Containers:** Utilise airtight containers or vacuum-sealed bags to safeguard food from humidity, vermin, and atmospheric contact.
  - **Labeling:** It is important to clearly mark all containers with the contents and expiration dates in order to make it easier to track and rotate them.
- **Inventory Management:**
  - **Regular Checks:** Perform routine inventory audits to oversee stock quantities and detect any goods that require utilisation or replacement.
  - **Organize by Expiration:** Arrange food items by expiration date, with the oldest items at the front to ensure they are used first.

3. Rotating Your Food Supplies

Implementing a methodical rotation procedure ensures the preservation and practicality of your food stockpile.

- **First-In, First-Out (FIFO) System:**
  - **Principle:** Employ the FIFO (First-In, First-Out) strategy to guarantee the utilisation of older items prior to newer ones. This aids in the prevention of food spoilage or wastage.
  - **Implementation:** Arrange your storage space in a manner where recently acquired items are positioned towards the rear, while older items are placed towards the front, facilitating convenient retrieval.

- **Meal Planning and Usage:**
  - **Incorporate Stockpiled Items:** Regularly include stored goods into your meal planning to ensure they are consumed before their expiration date. Develop recipes and menus that optimise the utilisation of your available resources.
  - **Adjust Inventory:** Revise your inventory list according to consumption trends and alterations in family requirements.
- **Monitoring Expiration Dates:**
  - **Track Dates:** Maintain a log of the expiration dates for all items in your stockpile. Create reminders to regularly inspect and utilise products before they reach their expiration dates.
  - **Replace and Replenish:** Substitute outdated or nearly expired products with new provisions to uphold the efficiency of your stockpile.
- **Handling Spoiled or Compromised Items:**
  - **Inspection:** Conduct routine examinations of your food supplies to detect any indications of deterioration, infestation, or harm. Dispose of any contaminated materials promptly.
  - **Cleaning:** Regularly clean and sanitise storage containers and spaces to prevent contamination and maintain food safety.

4. Updating and Expanding Your Reserves

Continuously adjusting your food stockpiles is crucial for maintaining long-term readiness and efficiency.

- **Review and Adjust:**
  - **Assess Needs:** Regularly evaluate your food stockpile and determine if your requirements have altered due to factors such as the number of family members, dietary choices, or developing circumstances.
  - **Expand as Necessary:** Expand or modify your inventory to fit evolving requirements, novel food products, or extended objectives.
- **Stay Informed:**
  - **New Products:** Stay updated on the latest advancements in long-term storage items or practices that can improve your food reserves..
  - **Emerging Risks:** Remain vigilant about possible hazards or alterations in food accessibility that could affect your approach to stockpiling.

Advice on Appropriate Shelf Life and Storage

Properly storing your food supplies is crucial for preserving their quality and prolonging their shelf life. This section offers pragmatic advice and optimal methods for efficient food preservation and the management of product longevity.

1. **Ideal Conditions for Storage**

   - **Temperature Control:**
     - **Cool Environment:** Preserve food in a temperature-controlled and consistent setting. The optimal temperature range for most dehydrated foods is 50°F to 70°F (10°C to 21°C).
     - **Avoid Heat Sources:** Avoid placing food near heat-emitting objects such as stoves, ovens, and direct sunlight. High temperatures can cause food to deteriorate and reduce its storage time.

   - **Humidity Control:**
     - **Low Humidity:** Strive for a low humidity setting to mitigate problems caused by moisture, such as mould, spoilage, and bugs. If needed, employ dehumidifiers or silica gel packs.
     - **Air Tightness:** Ensure that storage containers are securely sealed to avoid the ingress of moisture.

   - **Darkness:**
     - **Light Exposure:** To preserve the quality and prevent degradation of food, it is advisable to store it in a location devoid of light or in containers that do not allow light to pass through, as exposure to light can lead to nutrient depletion and spoiling.

2. **Choosing the Right Containers** •

   - **Airtight Containers:**
     - **Types:** Utilise hermetically sealed receptacles such as glass jars, plastic bins, or vacuum-sealed bags to safeguard food against the presence of air and moisture. Make sure they are tightly wrapped to preserve freshness.
     - **Size:** Select containers that are appropriate for the amount of food you have. To ensure adequate sealing and air removal, it is important to avoid overfilling.

   - **Vacuum Sealing:**
     - **Benefits:** Vacuum sealing eliminates air from bags, hence decreasing the likelihood of spoiling and prolonging the duration of freshness. Perfect for dehydrated goods and freeze-dried foods.
     - **Equipment:** Acquire a high-quality vacuum sealer and utilise vacuum-seal bags to efficiently preserve and store large quantities of products.

- **Mylar Bags and Oxygen Absorbers:**
  - **Use:** Mylar bags, when combined with oxygen absorbers, are highly effective for preserving grains, beans, and other dehydrated food items for extended periods of time. They aid in preserving the food by inhibiting the degradation caused by oxygen.
  - **Sealing:** Thermally bond Mylar bags are used to create an airtight seal, effectively preventing the entry of moisture and vermin.

3. Managing Shelf Life

- **Expiration Dates:**
  - **Check Regularly:** Regularly check the expiration dates of all food items and use them as a reference for when to rotate and consume them.
  - **Rotate Stock:** Apply the First-In, First-Out (FIFO) approach to prioritise the utilisation of older items above newer ones.

- **Shelf Life of Common Foods:**
  - **Canned Goods:** The usual duration is from 2 to 5 years, provided that it is stored appropriately. Prior to usage, inspect for any signs of swelling, rust, or leaks.
  - **Dry Beans and Rice:** The shelf life of the product might range from 1 to 2 years when stored in containers that are airtight. Prolong the duration of freshness by employing appropriate sealing and storing techniques.
  - **Freeze-Dried Foods:** If stored in Mylar bags or vacuum-sealed containers, the product can have a lifespan of 10-25 years.

- **Handling Spoiled or Compromised Items:**
  - **Inspection:** Conduct routine examinations of food to detect indications of decay, such as unpleasant odours, changes in colour, or abnormal consistency. Dispose of any contaminated materials promptly.
  - **Pest Control:** Employ mechanisms and maintain cleanliness in storage rooms to avert pest invasions.

4. Labeling and Tracking

- **Labeling:**
  - **Information:** Ensure that all containers are properly labelled with the contents, date of purchase, and expiration date. Utilise waterproof markers or labels to guarantee readability.
  - **Visibility:** Position labels in prominent locations to enhance rapid identification and streamline inventory management.

- **Inventory Management:**
    - **Record Keeping:** Keep a documented or electronic record of your food provisions. Regularly update it to monitor usage and determine when replacements are necessary.
    - **Checklists:** Utilise checklists to guarantee that all food items are accounted for and appropriately rotated.

5. Special Considerations

- **Food Preservation Techniques:**
    - **Canning:** Home canning is a technique used to preserve fruits, vegetables, and meats. Adhere to correct canning protocols and instructions to guarantee safety.
    - **Dehydrating:** Dehydrated foods possess an extended duration of storage and exhibit a low weight. To eliminate moisture from fruits, vegetables, and meats, employ a dehydrator or oven.
- **Emergency Food Storage:**
    - **Ready-to-Eat Meals:** Ensure the inclusion of emergency rations such as MREs (Meals Ready-to-Eat) for circumstances when cooking may not be practical.
    - **Water Storage:** Make sure you possess an ample quantity of uncontaminated water, as it is essential for the preparation of most dishes and for consumption.

CHAPTER 3
ENSURING ACCESS TO WATER SECURITY

Evaluating the Water Needs of Your Home

Comprehending and computing the water requirements of your residence is essential for efficient water security strategizing. This section will provide you with instructions on how to determine the amount of water your home needs and how to make necessary preparations.

1. **Calculating Your Daily Water Use**

   - **Basic Guidelines:**
     - **Daily Requirement:** Every individual need around 1 gallon (equivalent to 3.8 litres) of water daily to fulfil their demands for drinking, cooking, and sanitary purposes. Customise this according to individual requirements and specific weather conditions in the area.
     - **Family Size:** To get the total daily water demand, simply multiply the daily water requirement by the number of individuals in your household.
   - **Additional Uses:**
     - **Cooking:** Take into consideration the additional amount of water required for food preparation and cooking. Typically, the amount of liquid consumed by each individual on a daily basis might range from approximately 1-2 quarts (1-2 litres).
     - **Sanitation:** Add water for personal cleanliness, like washing your hands, brushing your teeth, and cleaning. This could mean that each person needs an extra 1.8 to 7.6 litres (1-2 gallons) of water every day.

2. **Calculating Long-Term Water Storage Needs**

   - **Short-Term Storage:**
     - **Initial Supply:** Plan to store enough water to last for at least two weeks in case of a disaster. To find out how much water is needed, multiply the daily needs by 14 days.

- **Long-Term Storage:**
  - **Extended Supply:** For longer-term plans, you might want to store enough water for one to three months. To find the total amount needed, multiply the daily needs by 30 to 90 days.
  - **Seasonal Adjustments:** Change how much room you need based on the time of year, like when you use more during hot weather.

3. **Evaluating the Sources of Water Supply**
   - **Assessing Existing Storage:**
     - **Current Reserves:** Conduct an assessment of all the water storage you currently own, which includes bottled water, stored barrels, and other containers.
     - **Storage Capacity:** Assess the overall capacity of your existing water storage and ascertain whether it satisfies your calculated requirements.
   - **Identifying Additional Sources:**
     - **Rainwater Collection:** Evaluate the feasibility of gathering rainwater and determine the storage capacity of your rain barrels or collection systems.
     - **Natural Sources:** Locate and assess the availability of adjacent natural water sources and evaluate their potential for use in emergency situations. Make sure you have purifying techniques accessible for these sources.

4. **Planning for Emergency Scenarios**
   - **Water Shortages:**
     - **Contingency Plans:** Create strategies for situations in which your main water sources are compromised or inaccessible. Take into account alternative sources for backup and employ purifying techniques.
     - **Emergency Kits:** Ensure the availability of clean water in emergency situations by include portable water filtration and purification devices in your emergency kits.
   - **Health and Safety Considerations:**
     - **Hydration Needs:** Take note of heightened need for fluid intake during periods of illness, elevated temperatures, or vigorous physical exertion. Revise the water storage plans to account for these concerns.
     - **Water Purity:** Maintain the cleanliness and safety of your stored water by adhering to appropriate storage and rotation procedures.

5. Regular Monitoring and Maintenance

- **Periodic Checks:**
  - **Storage Conditions:** Frequently inspect the state of your water storage containers and provisions. Search for indications of pollution or deterioration.
  - **Inventory Updates:** Regularly update your water inventory to accurately track usage, replacement, and any fluctuations in water requirements.

- **Rotation and Replacement:**
  - **Use and Replenish:** Ensure the rotation of your water supplies by utilising outdated inventory and replacing it with fresh water. This practice aids in preserving the quality of water and guarantees that your reserves are consistently prepared for use.

- **System Maintenance:**
  - **Inspect Systems:** To ensure the correct functioning of your water collection or purification systems during emergencies, it is important to frequently inspect and maintain them.

6. Planning for Special Situations

- **Family Members with Special Needs:**
  - **Medical Needs:** Take into account the water requirements of family members who have medical issues that may want extra hydration or specialised attention.
  - **Infants and Elderly:** Modify water reservoir management and strategic planning to accommodate the unique hydration needs of infants, elderly individuals, or anyone with specific hydration requirements.

- **Long-Term Sustainability:**
  - **Water Conservation:** Adopt water conservation tactics and techniques to decrease total usage and prolong the lifespan of your stored water.
  - **Community Resources:** Investigate community resources and networks to find extra assistance or water sources in the event of extended emergencies.

Solutions for Safe Water Storage

Ensuring the appropriate storage of water is crucial for preserving its safety and usability. This section offers detailed instructions for selecting and preserving efficient water storage options to guarantee access to uncontaminated and secure water in emergency situations.

1. Selecting the Right Containers

- **Food-Grade Containers:**

- **Material:** Utilise receptacles constructed from food-safe plastic or glass. These materials are non-toxic and do not release any dangerous substances when used to store drinking water.
- **Types:** Available choices consist of plastic barrels that are free from BPA, jars made of glass, and tanks constructed from stainless steel. Refrain from utilising containers that have previously contained non-edible substances.

- **Water Storage Barrels:**
  - **Size and Capacity:** Choose barrels with capacities ranging from 15 to 55 gallons (56 to 208 litres) based on the amount of storage you require and the space you have.
  - **Features:** Search for barrels that have a securely sealed lid and a spigot for convenient distribution. Certain barrels are equipped with integrated UV protection to inhibit the growth of algae.

- **Portable Water Containers:**
  - **Sizes:** Containers of a smaller capacity, ranging from 1 to 5 gallons, are beneficial for storing items for a brief period and for easy transportation. Perfect for inclusion in emergency kits or for regular consumption.
  - **Durability:** Select resilient, foldable water containers for convenient storage and transportation. Ensure that the materials used are safe for contact with food.

2. Proper Storage Techniques

- **Cleanliness:**
  - **Sanitize Containers:** Prior to storing water, meticulously cleanse and disinfect all containers using a solution of water and unscented bleach (1 tablespoon of bleach per gallon of water). Thoroughly rinse and allow to dry naturally.
  - **Avoid Contamination:** Make sure that the containers are totally dry prior to filling them. Ensure that the items are stored in a hygienic setting to avoid any form of infection.

- **Sealing and Protection:**
  - **Airtight Sealing:** Ensure that all containers are securely sealed to prevent any contamination from external sources. Utilise airtight lids or caps to prevent the entry of dust, bugs, and pollutants.
  - **Protection from Light:** To minimise the formation of algae and degradation caused by exposure to light, it is advisable to store water containers in a cool and dark location. Utilise containers that are not transparent, if feasible.

- **Temperature Control:**
  - **Cool Environment:** Preserve the freshness of water by storing it in a cool environment. Steer clear of regions with elevated temperatures or direct exposure to sunshine.
  - **Avoid Freezing:** When storing water in cold climates, it is important to either insulate the containers or keep them in areas where freezing does not occur. Freezing can cause damage to the containers and perhaps affect the quality of the water.

3. **Rotating and Managing Water Supplies**
   - **Rotation Schedule:**
     - **Regular Use:** Develop a First-In, First-Out (FIFO) mechanism that prioritises the utilisation of older water supplies over newer ones. This helps ensuring that water is consumed during its ideal storage duration.
     - **Replenishment:** Periodically replace the stored water every 6-12 months, or as advised by storage rules, to uphold its quality.
   - **Labeling:**
     - **Identification:** Ensure that every water container is clearly marked with the date of storage and other pertinent details. Utilise waterproof markers or adhesive labels to guarantee legibility.
     - **Tracking:** Keep a record of the amount of water kept and the timeline for replenishing it to effectively monitor your stockpile.

4. **Alternative Water Storage Methods**
   - **Rainwater Harvesting:**
     - **Collection Systems:** Install rain barrels or cisterns to gather and retain rainwater. Make sure that they have screens or filters in place to prevent the entry of trash and insects.
     - **Storage Capacity:** Take into account the dimensions of your collection system in relation to the precipitation patterns in your area and the water requirements of your family.
   - **Emergency Water Pouches:**
     - **Single-Use Pouches:** Emergency water pouches are small in size and have a prolonged shelf life. They are optimal for inclusion in emergency kits and for instances where there is limited space.
     - **Usage:** Verify the expiration dates and substitute any pouches that are necessary in order to guarantee their safety for eating.

5. **Water Purification and Treatment**
   - **Pre-Treatment:**
     - **Purification Solutions:** Utilise water purification tablets or solutions to process water before to storage if obtained from uncertain or potentially contaminated sources.
     - **Boiling:** Prior to storage, it is advisable to heat water to its boiling point in order to eliminate harmful microorganisms and guarantee its safety in the absence of other purifying measures.
   - **Post-Storage Purification:**
     - **Testing:** Regularly conduct tests on stored water for pollutants, particularly if the storage conditions are uncertain. Utilise home water testing kits to evaluate the quality of water.
     - **Re-Treatment:** Prior to usage, cleanse water by suitable techniques (such as filtration, boiling, or chemical treatments) in the event that contamination is identified.

6. **Safety Considerations**
   - **Pest Prevention:**
     - **Insect Control:** Utilise screens or protective covers to hinder insects from gaining access to water storage. Regularly inspect containers for indications of pest infestation.
     - **Rodent Proofing:** Place containers in locations that are less easily reached by rodents and inspect for any indications of contamination or harm.
   - **Disposal of Contaminated Water:**
     - **Proper Disposal:** In the event that water becomes polluted or exhibits indications of spoiling, it is imperative to appropriately dispose of it and fully cleanse the containers before reusing them.
     - **Preventing Contamination:** Maintain cleanliness and sanitation of all containers and storage locations to avoid potential contamination in the future.

Methods of Filtration and Purification

It is crucial to preserve the purity of your water in order to uphold health and safety standards during emergency situations. This section discusses a range of filtration and purification techniques to ensure that your water is free from contaminants and suitable for consumption.

1. **Boiling**
   - **Procedure:**
     - **Boil Time:** To effectively eliminate the majority of pathogens, such as bacteria, viruses, and protozoa, it is necessary to bring water to a vigorous boil for a minimum of 1 minute (3 minutes at altitudes exceeding 6,500 feet or 2,000 meters).
     - **Containers:** Utilise pots or kettles that are free from dirt or impurities. Refrain from boiling substances in containers composed of materials that have the potential to release chemicals, such as specific types of plastics.
   - **Limitations:**
     - **Chemical Contaminants:** Boiling does not eliminate chemical pollutants or enhance flavour.
     - **Energy Requirement:** Dependent on a heat source, which may be inaccessible in some circumstances.

2. **Filtration**
   - **Activated Carbon Filters:**
     - **Function:** Eliminates chlorine, sediment, and certain organic pollutants. Enhances flavour and aroma.
     - **Types:** Comprises of pitcher filters, faucet-mounted filters, and under-sink systems.
     - **Maintenance:** Periodically replace filter cartridges according to the manufacturer's specifications to ensure optimal efficacy.
   - **Ceramic Filters:**
     - **Function:** Filters out bacteria, protozoa, and sediment. Often used in portable water filters.
     - **Types:** Comprises of gravity-fed filters, pump filters, and countertop systems.
     - **Maintenance:** Regularly clean the ceramic element to ensure optimal flow rate and efficacy.
   - **Reverse Osmosis (RO) Systems:**
     - **Function:** Eliminates a broad spectrum of impurities, such as salts, chemicals, and heavy metals.
     - **Components:** Usually consists of a preliminary filter, reverse osmosis membrane, and final filter.
     - **Maintenance:** Follow the manufacturer's recommendations and replace filters and membranes accordingly. Perform routine inspections to detect any leaks or performance deficiencies.
   - **Ultrafiltration (UF) Systems:**
     - **Function:** Removes germs, viruses, and bigger particles by filtration. Does not eliminate or extract dissolved substances.

- **Types:** Encompasses portable filters, systems installed under the sink, and devices placed on the countertop.
- **Maintenance:** Follow the directions to clean and change filters.

3. **Chemical Purification**
   - **Chlorine Tablets:**
     - **Function:** Exhibits bactericidal, virucidal, and protozoic properties. Frequently employed for the purpose of emergency water treatment.
     - **Usage:** Adhere to the dosage and contact time specified by the manufacturer's instructions. Typically, the recommended dosage is 1 pill for every 2 litres (0.5 gallons) of water.
     - **Considerations:** Chlorine has the potential to impact the flavour and may not be appropriate for extended periods of use.
   - **Iodine Tablets:**
     - **Function:** Potent against germs and viruses. Utilised during critical and life-threatening circumstances.
     - **Usage:** Adhere to the prescribed dose guidelines. Generally, the recommended dosage is 1 tablet for every litre (equivalent to 0.26 gallons) of water.
     - **Considerations:** Pregnant women and individuals with thyroid disorders should avoid using iodine. Additionally, it has the potential to impact one's sense of taste and may not be suitable for extended periods of use.
   - **Bleach:**
     - **Function:** Kills bacteria, viruses, and some protozoa. Use unscented household bleach with a 5-6% sodium hypochlorite concentration.
     - **Usage:** To treat clean water, add 2 drops of bleach per litre (0.26 gallons). For cloudy water, add 4 drops per litre. Allow it to rest for a minimum of 30 minutes prior to usage.
     - **Considerations:** Exercise caution with excessive usage, as it may result in the accumulation of detrimental chemical levels.

4. **UV Purification**
   - **UV Sterilizers:**
     - **Function:** Utilises ultraviolet radiation to eradicate bacteria, viruses, and protozoa. Efficient in treating water that is free from impurities.

- **Types:** Encompasses portable devices, ultraviolet pens, and more substantial systems designed for residential applications.
- **Usage:** Adhere to the guidelines provided by the manufacturer for operating and maintaining the product. To achieve optimal effectiveness, it is important to ensure that the water is clear.

- **Advantages:**
    - **Speed:** Offers rapid filtration while causing minimal alterations in taste or odour.
    - **Effectiveness:** Efficient in combating a wide spectrum of bacteria.

5. Distillation

- **Procedure:**
    - **Process:** Heat water until it reaches its boiling point to produce steam, and then cool the steam until it turns back into liquid water, separating impurities in the process.
    - **Types:** Includes home distillation units and improvised setups.
    - **Advantages:** Eliminates a wide range of impurities, such as salts, heavy metals, and compounds.

- **Limitations:**
    - **Energy Intensive:** Dependent on a heat source and has a tendency to be time-consuming.
    - **Cost:** Domestic distillation apparatuses can be costly and necessitate frequent upkeep.

6. Practical Tips for Filtration and Purification

- **Pre-Filtration:**
    - **Sediment Removal:** Employ pre-filters or sediment filters to eliminate sizable particles prior to employing more refined filtration techniques.
    - **Pre-Treatment:** Apply a coagulant to turbid or polluted water in order to precipitate particles prior to filtration.

- **Combining Methods:**
    - **Multi-Stage Purification:** Employ various techniques, such as employing a filter followed by chemical treatment or UV purification, to increase the level of safety.
    - **Adaptation:** Select the suitable combination of techniques according to the water's quality, accessibility, and emergency circumstances.

- **Storage and Handling:**
  - **Clean Containers:** Ensure that you consistently utilise sterile containers while storing cleansed water to prevent any potential reintroduction of contaminants.
  - **Testing:** Regularly assess the quality of water, if feasible, particularly if it has been held for a prolonged duration.

7. Safety Considerations

- **Understanding Limitations:**
  - **Filter Lifespan:** Consistently replace filters and adhere to maintenance requirements to guarantee ongoing efficiency.
  - **Contaminants:** Take into account the constraints of each approach and employ numerous techniques if required to achieve the highest level of security.

- **Emergency Preparedness:**
  - **Portable Solutions:** Ensure that your emergency preparedness packs are equipped with portable filtration and purification devices.
  - **Training:** Acquaint yourself with the functioning of purifying equipment and techniques to ensure proficient utilisation when necessary.

This section offers an in-depth exploration of different filtration and purification techniques, including their purposes, benefits, drawbacks, and practical recommendations for achieving water that is both clean and safe. It aids readers in comprehending and executing the most efficient water purifying techniques for emergency circumstances.

CHAPTER 4
CREATING YOUR OWN POWER

Energy Requirements in an Emergency

Comprehending and controlling your energy requirements during a crisis is crucial for sustaining vital operations and guaranteeing survival. This section examines the process of evaluating and ranking energy requirements, adjusting to scarce resources, and executing tactics to maximise energy efficiency.

1. **Assessing Energy Requirements**

   - **Identifying Critical Systems:**

     - **Essential Equipment:** Identify the essential systems and equipment necessary for survival, including refrigeration, heating, lighting, medical devices, and communication gadgets.

     - **Prioritization:** Arrange these systems in order of their significance for urgent necessities and long-term viability. Concentrate on the aspects that are crucial for maintaining good health, ensuring safety, and promoting overall well-being.

   - **Calculating Energy Consumption:**

     - **Wattage and Usage:** Determine the power consumption of every essential device and approximate the duration of their usage in order to assess the energy needs. As an illustration, a refrigerator may utilise 200 watts of power and operate for 8 hours daily, resulting in a total of 1,600 watt-hours each day.

     - **Daily Needs:** To calculate your entire daily energy usage in watt-hours (Wh) or kilowatt-hours (kWh), add up the energy requirements of all necessary equipment.

2. **Planning for Limited Power Resources**

   - **Load Shedding:**

     - **Prioritize Usage:** Enforce load shedding by activating only the most essential devices during periods of restricted power supply. Temporarily deactivate non-essential equipment to preserve energy.

- **Scheduling:** Create a timetable for utilising vital equipment to guarantee they are energised, when necessary, while also prolonging the lifespan of your power source.

- **Energy Efficiency:**
    - **Energy-Saving Devices:** Utilise energy-efficient equipment and lighting in order to decrease overall energy use. As an illustration, LED lamps utilise a lower amount of energy compared to incandescent bulbs.
    - **Insulation and Conservation:** Enhance the insulation of your home to decrease the amount of energy used for heating and cooling. To regulate indoor temperature, it is advisable to close curtains, apply weather stripping, and plug any leaks.

3. Adapting to Power Outages

- **Backup Power Solutions:**
    - **Generators:** Employ backup generators to provide electricity during periods of power failure. Select a generator size that aligns with your energy requirements and guarantees proper maintenance and fuelling.
    - **Battery Storage:** Utilise battery systems to store energy for the purpose of utilising it during power disruptions. Choose batteries that have a capacity that is adequate to fulfil your essential power requirements.

- **Alternative Energy Sources:**
    - **Solar and Wind:** If possible, utilise solar or wind energy to augment electricity requirements. Make sure you have the required equipment and storage systems established.
    - **Hydroelectric and Thermoelectric:** If you have access to appropriate resources, consider investigating the potential of small-scale hydropower or thermoelectric systems.

4. Emergency Energy Conservation Strategies

- **Efficient Use of Power:**
    - **Power Management:** Utilise power management systems for the purpose of monitoring and regulating energy consumption. Give priority to crucial devices and reduce power consumption during periods of high demand.
    - **Smart Practices:** Adopt energy-conservation measures, such as switching off lights when not in use, disconnecting unused devices from power sources, and utilising programmable thermostats.

- **Alternative Lighting Solutions:**
  - **Battery-Powered Lights:** Use battery-powered lanterns and flashlights for lighting. Consider rechargeable options to reduce waste and ensure availability during outages.
  - **Candles and Oil Lamps:** As a provisional measure, utilise candles or oil lamps. Exercise prudence to mitigate the risk of fire dangers and provide adequate ventilation.
- **Heating and Cooling:**
  - **Temporary Solutions:** Utilise alternative heating techniques, such as portable heaters or wood stoves, in the absence of conventional heating systems. To achieve cooling, employ fans or open windows to provide proper ventilation.
  - **Insulation:** Enhance the insulation of your home to effectively preserve heat or cold temperatures. Utilise blankets, thermal curtains, and draft stoppers to optimise comfort.

5. Monitoring and Adjusting Power Usage

- **Energy Monitoring Tools:**
  - **Usage Meters:** Utilise energy meters to monitor the power use of individual devices as well as the general usage within the household. This information facilitates the efficient management and adaptation of energy consumption.
  - **Smart Devices:** Integrate intelligent plugs or outlets to remotely monitor and regulate energy consumption. Establish notifications for excessive usage to proactively monitor and control consumption.
- **Adjusting Based on Conditions:**
  - **Power Availability:** Modify energy consumption according to the presence of electricity from generators, batteries, or other alternate sources. Adapt resource management to suit the present circumstances.
  - **Emergency Scenarios:** Anticipate different possibilities and have backup strategies. For instance, prepare for prolonged periods without electricity or situations in which the availability of power may be restricted.

6. Safety Considerations

- **Electrical Safety:**
  - **Proper Installation:** Ensure proper installation and maintenance of all electrical systems, such as generators and battery storage, in accordance with safety rules. Prevent the excessive load on electrical circuits and ensure the adoption of suitable wiring.
  - **Ventilation:** Run generators and other fuel-powered machines in well-ventilated places to keep carbon monoxide from building up. Be sure to follow safety rules when you store and use fuel.

- **Fuel Management:**
  - **Storage:** Ensure proper storage and adherence to laws when handling fuel. Store fuel in containers that have been approved for this purpose and ensure that they are kept at a safe distance from any potential sources of fire.
  - **Conservation:** Optimise fuel use and ensure a sufficient reserve for prolonged periods without power. Develop and execute tactics to preserve fuel and prolong its utilisation.

This comprehensive part provides readers with a thorough understanding of how to effectively address and control their energy requirements in times of need. The content encompasses the evaluation of energy needs, adjusting to power interruptions, applying methods to conserve energy, and ensuring safety when utilising alternative power sources.

Off-Grid Power Options: Generators, Wind, and Solar

Off-grid power solutions are necessary when traditional power sources are not accessible. This section examines three primary techniques for producing electricity autonomously: solar power, wind power, and generators. Every approach possesses unique advantages, constraints, and factors to consider when implementing it effectively.

1. Solar Power Solutions

- **Solar Panels:**
  - **Types and Efficiency:**
    - **Monocrystalline Panels:** Renowned for its exceptional efficiency and compact design, making it perfect for constrained spaces.
    - **Polycrystalline Panels:** Marginally less efficient but more economical.
    - **Thin-Film Panels:** Characterised by its low weight and flexibility, this product is well-suited for particular use cases, however it may not be as efficient in general.
  - **Installation:**
    - **Site Assessment:** Arrange the panels in a way that optimises their exposure to sunlight. Mount on rooftops, ground mounts, or portable configurations.
    - **Mounting Options:** Utilise adjustable mounts to incline panels for maximum sunlight absorption.

- **Solar Battery Storage:**
  - ➤ Types:
    - **Lead-Acid Batteries:** Although affordable, these items necessitate frequent maintenance and have a shorter lifespan.
    - **Lithium-Ion Batteries:** Increased efficiency and extended lifespan, albeit at a higher cost.
    - **Gel Batteries:** This product requires little maintenance and performs well in a wide range of temperatures.
- **Capacity Planning:**
  - **Determine Storage Needs:** Determine the necessary battery capacity by considering the amount of energy consumed on a daily basis and the desired length of time the battery should provide backup power.
  - **Integration:** Integrate batteries into the solar system to store surplus energy produced during periods of abundant sunlight.
- **Inverters:**
  - ➤ Function:
    - **DC to AC Conversion:** Transform the direct current (DC) generated by solar panels into the alternating current (AC) that is utilised by household appliances.
  - ➤ Types:
    - **Pure Sine Wave Inverters:** Deliver reliable and consistent electricity that is free from impurities, making it appropriate for delicate electrical devices.
    - **Modified Sine Wave Inverters:** More affordable yet may result in interference with some equipment.
    - **Hybrid Inverters:** Integrate solar and battery management capabilities to enhance energy utilisation and storage efficiency.

2. Wind Power Solutions

- **Wind Turbines:**
  - ➤ Types:
    - **Horizontal-Axis Turbines:** Most prevalent, well-suited for regions with strong winds and a significant land area.
    - **Vertical-Axis Turbines:** Compact and ideal for urban or small-scale use.

- **Installation:**
  - **Location:** Install at locations characterised by a steady and uninterrupted flow of wind, ensuring that there are no obstacles such as trees or buildings nearby.
  - **Height:** Place the turbines at an elevation that allows them to harness the most favourable wind velocities while minimising turbulence.

- **Wind Battery Storage:**
  - **Integration:**
    - **Storage System:** Link wind turbines to battery storage devices to store generated electricity for utilisation during periods of low wind conditions.
    - **Capacity Planning:** Evaluate the necessary battery capacity by considering the turbine's power output and your energy storage requirements.
  - **Maintenance:**
    - **Regular Checks:** Examine the blades, gearboxes, and electrical components for signs of damage or deterioration. Conduct regular maintenance to provide the best possible performance.
    - **Safety:** To minimise accidents and maintain safe operation, it is important to ensure appropriate installation and grounding.

3. Generators

- **Types of Generators:**
- **Portable Generators:**
  - **Fuel Options:** Usually fuelled by gasoline, propane, or diesel. Appropriate for temporary or urgent utilisation.
  - **Capacity:** Select a generator that aligns with your specific power requirements. Portable generators are offered in a range of wattages.
- **Standby Generators:**
  - **Automatic Operation:** Permanently installed systems that initiate automatically in the event of power failures. Frequently fuelled by natural gas or propane.
  - **Capacity and Integration:** Designed to accommodate entire household or essential power demands. Professional installation and connection to the home's electrical infrastructure are necessary.
- **Fuel Management:**

- **Storage:** Keep fuel in safe places and in containers that have been cleared. Follow the rules in your area for storing and handling fuel.
- **Conservation:** Generators should be used properly to save fuel. Put the most important items first and don't overload the generator.

- **Usage Tips:**
  - **Load Management:** Effectively regulate the electrical demand to prevent excessive strain on the generator. Employ a transfer switch to securely link the generator to the electrical system of your residence.
  - **Ventilation:** To avoid the accumulation of carbon monoxide, it is important to use generators in places that are properly ventilated. Adhere to safety protocols regarding the positioning and utilisation of the item.

4. **Hybrid Systems**
   - **Combining Technologies:**
     - **Solar and Wind Integration:** Integrate solar panels and wind turbines to establish a more dependable and uniform electricity source. Each system has the ability to enhance the strengths and mitigate the shortcomings of the other.
     - **Backup Generators:** Utilise generators as an auxiliary power supply to complement solar and wind systems during prolonged periods of reduced energy generation.
   - **System Design:**
     - **Energy Management:** Develop and deploy a power management system to oversee and regulate the transfer of energy among solar, wind, and generator sources.
     - **Storage Solutions:** Incorporate battery storage to accumulate surplus energy generated by solar or wind systems and utilise it during periods of reduced production.

5. **Practical Considerations**
   - **Budget and Costs:**
     - **Initial Investment:** Take into account the initial expenses associated with acquiring and setting up solar panels, wind turbines, and generators.
     - **Ongoing Maintenance:** Consider the expenses associated with the upkeep and maintenance of power systems. Consistent maintenance guarantees durability and dependable functionality.

- **Local Regulations:**
  - **Permits and Codes:** Ensure compliance with local rules and acquire the requisite permits for the installation of solar panels, wind turbines, or generators. Guarantee adherence to building codes and safety norms.
- **Scalability:**
  - **Future Expansion:** Develop a strategic plan to anticipate and accommodate anticipated future growth of your power infrastructure. It is advisable to increase the number of panels, turbines, or battery capacity as the requirements expand.

This section offers a thorough and inclusive examination of off-grid power alternatives, encompassing solar, wind, and generator systems. The document encompasses pragmatic factors for the installation, upkeep, and incorporation of systems to guarantee a dependable and enduring power provision in times of crisis.

Safety and Fuel Storage Issues

Ensuring proper fuel storage and safety measures are of utmost importance to guarantee dependable power generation and minimise potential hazards in emergency situations. This section provides guidelines for the proper storage of different types of fuel, including safety measures and regulatory requirements.

1. Fuel Types and Storage Options

- **Gasoline:**
  - **Storage Containers:** Utilise authorised, high-quality receptacles for petrol. Containers must be constructed from high-density polyethylene (HDPE) or metal and clearly labelled.
  - **Quantity Limits:** Do not exceed a storage capacity of 25 gallons of petrol in residential zones. Local regulations may impose restrictions on excessive quantities.
- **Propane:**
  - **Storage Tanks:** Utilise propane tanks that have been officially certified for storing purposes. Tanks are available in a range of sizes, spanning from small cylinders to enormous bulk tanks.
  - **Placement:** Properly position propane tanks outside in regions with good airflow, keeping them at a safe distance from structures, sources of ignition, and substances that can easily catch fire.
- **Diesel:**
  - **Storage Containers:** Diesel should be kept in approved barrels or tanks made for fuel. Diesel fuel tanks should have holes in them so that gas doesn't build up.

- **Stability:** Diesel is less volatile than gasoline, but proper storage is still essential to prevent contamination and degradation.
- **Alternative Fuels (e.g., Kerosene, Biofuels):**
  - **Special Containers:** Use containers that are made to hold certain types of fuel, like biofuels or oil. Follow the storage and handling instructions given by the maker.
  - **Shelf Life:** Know how long alternative fuels last and how they break down over time. Check and change old fuel on a regular basis to get the best performance.

2. Safety Considerations

- **Ventilation:**
  - **Proper Airflow:** Ensure that fuels are stored in areas with sufficient ventilation to avoid the accumulation of vapours, which may result in explosions or pose health risks. Refrain from storing fuels in confined or inadequately ventilated areas.
- **Temperature Control:**
  - **Avoid Extreme Temperatures:** Ensure that fuels are stored in surroundings with consistent and unchanging temperatures. Minimise contact with elevated temperatures or direct sunshine, as this can heighten the likelihood of evaporation and ignite.
- **Fire Safety:**
  - **Distance from Ignition Sources:** Ensure that flammable substances are kept at a safe distance from open flames, sparks, and electrical equipment. Ensure that you keep a sufficient distance from any sources that have the potential to cause ignite.
  - **Fire Extinguishers:** Place appropriate fire extinguishers near fuel storage areas. Ensure that the extinguishers are suitable for the type of fuel stored (e.g., Class B for flammable liquids).
- **Leak Prevention:**
  - **Container Integrity:** Conduct routine examinations of storage containers to identify any signs of leakage, fractures, or impairment. Substitute or mend containers that exhibit indications of deterioration.
  - **Spill Response:** Keep spill tools and absorbent materials on hand so that you can quickly clean up any spills that happen. Follow the right steps to get rid of contaminated items.

3. Regulations and Best Practices

- **Local Regulations:**
  - **Compliance:** Comply with the local legislation and recommendations for the storage of gasoline. Regulations can differ depending on the geographical area, the type of fuel, and the amount stored.

- **Permits:** Obtain necessary permits for large-scale fuel storage, especially for propane tanks and bulk fuel systems.

- **Labeling and Documentation:**
  - **Clear Labels:** It is essential to clearly designate all gasoline containers with the specific type of fuel and any cautionary notices regarding potential dangers. Effective labelling facilitates the management and prompt reaction to emergencies.
  - **Record Keeping:** Keep detailed records of fuel acquisitions, storage sites, and safety evaluations. Recording this data guarantees adherence to regulations and facilitates monitoring of fuel consumption.

- **Security Measures:**
  - **Access Control:** Limit entry to fuel storage locations exclusively to authorised workers. Utilise locks or security systems as a means of thwarting unauthorised entry or tampering.
  - **Emergency Plans:** Create and convey emergency protocols for fuel-related situations, encompassing guidelines for evacuations, spills, and fire management.

4. Fuel Management and Rotation

- **Inventory Management:**
  - **Track Usage:** Monitor and record fuel levels and consumption. Consistently oversee inventory to guarantee a sufficient quantity without excessive stockpiling.
  - **Rotation:** Implement a first-in, first-out (FIFO) strategy to prioritise the consumption of older fuel over newer fuel. This approach aids in the prevention of fuel degradation and ensures the maintenance of its quality.

- **Testing and Quality Control:**
  - **Check for Contamination:** Regularly conduct tests on stored fuel to detect any contamination or deterioration. Utilise gasoline additives as needed to uphold the standard and efficiency.
  - **Replace Expired Fuel:** To prevent potential issues with power generation equipment, it is important to regularly replace fuel that has expired or deteriorated.

This part offers a thorough examination of gasoline storage and safety considerations, encompassing various fuel kinds, safety measures, adherence to regulations, and optimal strategies for fuel supply management and maintenance.

CHAPTER 5
TECHNIQUES FOR HOME DÉFENSE

Defending Your House: Windows, Doors, and the Exterior

Ensuring the security of the access points and perimeter of your property is vital for a successful defence. This section offers comprehensive tactics for fortifying doors, windows, and the boundaries of your premises to bolster security and discourage prospective trespassers.

1. Reinforcing Doors

- **Solid-Core Doors:**

    - **Material:** Substitute hollow-core doors with either solid-core or metal doors. These products offer increased resistance to unauthorised access and provide improved security.

    - **Installation:** Ensure that doors are installed correctly, with minimal gaps around the edges. Utilise professional services if required for optimal installation.

- **High-Security Locks:**

    - **Deadbolts:** Ensure that deadbolt locks are installed with a minimum throw of 1 inch. Take into account high-security deadbolts that incorporate anti-picking and anti-bumping characteristics.

    - **Locksets:** Utilise robust locksets with strengthened strike plates. Make sure that all external doors are equipped with high-quality locks and that keys are stored in a secure manner.

- **Door Reinforcement:**

    - **Door Braces:** Mount door reinforcements or security grilles on doors, particularly on sliding doors or doors equipped with glass panels. These gadgets serve to impede the act of forcefully kicking or prying open a door.

    - **Reinforced Hinges:** Utilise robust hinges including pins that cannot be removed. It is advisable to attach hinge bolts to deter the removal of the door from its frame.

2. **Securing Windows**
   - **Window Locks:**
     - **Lock Types:** Install premium window locks, such as key-operated locks or sliding window locks, to ensure high-quality security. Make sure that all windows that can be easily reached are fitted with locks.
     - **Secondary Locks:** It is advisable to incorporate supplementary security measures, such as window bars or security pins, to enhance the level of protection.
   - **Shatterproof Films and Bars:**
     - **Shatterproof Film:** Install shatterproof or impact-resistant film on windows. This film enhances the durability of glass, making it more resistant to breakage and incursions.
     - **Window Bars:** Mount security bars or grills on windows that are susceptible to break-ins. Select designs that are both secure and visually appealing and verify that they can be readily opened from the interior in the event of an emergency.
   - **Window Sensors:**
     - **Alarm Sensors:** Place alarm sensors on windows to identify any unauthorised openings. Incorporate these sensors into your home security system to receive immediate notifications.

3. **Enhancing Perimeter Security**
   - **Fencing:**
     - **Perimeter Fencing:** Construct durable fencing at the boundary of your premises. Select materials such as chain-link, wood, or wrought iron based on your specific security requirements and desired visual appeal.
     - **Height and Design:** Select fencing that is of sufficient height to discourage climbing. If deemed suitable for your location, it is advisable to enhance security measures by using barbed wire or electric fencing.
   - **Gates:**
     - **Secure Gates:** Utilise premium-grade locks on gates and ensure they are adequately fastened. It is advisable to utilise gate operators equipped with remote access in order to manage the admission and exit of individuals.
     - **Gate Reinforcement:** Enhance the strength of gates by adding supplementary locking mechanisms or security bars. Ensure that the gates are robust and impervious to unauthorised access.

- **Landscaping:**
  - **Visibility:** Ensure proper upkeep of the landscaping to prevent the formation of concealed areas that could be used by trespassers. Maintain the pruning of shrubs and trees to guarantee unobstructed lines of vision surrounding your residence.
  - **Defensive Landscaping:** Utilise prickly vegetation or thick foliage in close proximity to windows and doors to establish organic obstacles. This might act as a deterrent to burglars and offer extra security.
- **Lighting:**
  - **Exterior Lighting:** Place motion-activated lights at strategic locations around the exterior of your house. Establish well-illuminated spaces to discourage trespassers and enhance nighttime visibility.
  - **Spotlights:** Utilise spotlights to illuminate crucial access points and areas with low light intensity. Ensure that the placement of lighting adequately covers all locations that are susceptible to potential harm or danger.

4. Security Technology

- **Surveillance Cameras:**
  - **Camera Placement:** Arrange the security cameras strategically to encompass all access points, such as doors, windows, and the surrounding regions. Make sure that the cameras are clearly visible in order to discourage potential offenders.
  - **Camera Features:** Select cameras that possess attributes such as superior resolution, nocturnal visibility capabilities, and the ability to detect motion. Integrate cameras with a centralised monitoring system to enable live surveillance.
- **Alarm Systems:**
  - **Motion Detectors:** Place motion sensors strategically along the boundaries of your residence to identify any abnormal movements. Incorporate these sensors into your alarm system to receive immediate notifications.
  - **Alarm Integration:** Make sure that your alarm system is linked to your security cameras and other sensors to create a comprehensive security solution.
- **Smart Home Integration:**
  - **Remote Access:** Utilise smart home technology to remotely manage and supervise security systems. Control camera feeds, alerts, and lights with your smartphone or tablet.
  - ○**Automated Alerts:** Establish automated notifications for all instances of security breaches or system faults. Personalise notifications to guarantee timely replies.

This section offers a thorough advice on reinforcing doors, windows, and the boundaries of your residence. By applying these tactics, you may greatly improve your home's security and provide better protection for your family and belongings in times of emergency.

Tools and Techniques for Defense

When being ready for a possible disaster, it is crucial to have the appropriate defensive tools and procedures available to safeguard yourself, your family, and your property. This section provides comprehensive information on several defensive tools, including their practical uses and strategies for maximising their effectiveness.

1. **Personal Defensive Tools**

   - **Pepper Spray:**
     - **Usage:** Pepper spray is a non-lethal device that temporarily disables an attacker by delivering severe discomfort to their eyes, skin, and respiratory system.
     - **Application:** Ensure that you keep pepper spray in a convenient place where it can be easily reached, and make sure you are well-acquainted with how to use it. To optimise efficiency, direct your aim at the assailant's face and release the spray in brief, intermittent bursts.

   - **Personal Alarms:**
     - **Function:** Personal alarms are devices that produce a high-decibel sound when activated. They are specifically designed to discourage potential intruders and notify nearby individuals of your emergency situation.
     - **Usage:** Keep an alarm on your keychain or in your bag at all times. Turn it on if you feel threatened or need to get people's attention to an emergency.

   - **Tactical Flashlights:**
     - **Features:** Tactical flashlights give off a lot of light and can be used to protect yourself by briefly blinding an attacker.
     - **Usage:** Use the torch to light up dark places and throw off people who might be trying to break in. Pick types that can flash to scare people away even more.

   - **Self-Defense Weapons:**
     - **Batons:** For self-defence, you can use an expandable stick to knock out an attacker. Practise using a stick to make sure you know how to handle it well.
     - **Knives:** You should only carry a knife if you know how to use it and what the law says about it. When your life is in danger, you can use a knife to protect yourself.

2. Home Defense Tools

- **Security Bars and Grills:**

    - **Installation:** Put grilles or security bars on windows and other easy-to-crack entry places. Pick things that will last and can't be changed.

    - **Usage:** For emergency doors, make sure that security bars and grills are easy to open from the inside.

- **Door Jammers and Barricades:**

    - **Door Jammers:** Door jammers are portable gadgets that make doors stronger and stop people from breaking in. Put them under the door handle to keep the door closed.

    - **Barricades:** To make doors stronger from the inside, use door bars or security bars. Intruders may not be able to get in through these methods.

- **Safe Rooms and Panic Rooms:**

    - **Construction:** Set aside or build a safe room in your home that is stocked with important items and ways to communicate.

    - **Usage:** Use the safe room as a safe place to be in case of an emergency or an attacker. Make sure it is strong and has things like food, a phone, and a first-aid kit.

3. Defensive Techniques

- **Situational Awareness:**

    - **Observation:** Check your surroundings often and be on the lookout for anything strange. Pay attention to how things around you or the way people act change.

    - **Alertness:** Stay aware, especially when you're in a dangerous place or one you don't know. If you feel like something is off, trust your gut and take extra care.

- **Evacuation Plans:**

    - **Planning:** Make escape plans for different situations and practise them. Make sure that everyone in the family knows how to leave and where to go.

    - **Practice:** Do drills on a regular basis to get used to leaving your home. If something goes wrong, make sure everyone knows where to go and what to do.

- **Self-Defense Training:**

    - **Techniques:** Learn simple self-defence moves, like how to fight back against common strikes and get out of holds. You might want to take self-defence classes to get more complete training.

- **Conditioning:** Regularly train and improve your self-defence skills to stay skilled. You can use hitting bags or defensive training gear to help you get better.
- **Home Defense Drills:**
  - **Scenarios:** Do practice for home defence to act out different types of threats. Practise locking up your house, using safety tools and what to do if someone breaks in.
  - **Family Involvement:** Family members should all take part in drills to make sure everyone knows what to do in an emergency.

4. Legal and Ethical Considerations

- **Legal Compliance:**
  - **Weapon Laws:** Find out what the rules are in your area about having and using defence tools. Make sure that legal standards are met to avoid legal problems.
  - **Use of Force:** Learning about the rules that govern using force in self-defence is important. When you need to, and only when the threat warrants it, use defence tools.
- **Ethical Considerations:**
  - **Proportional Response:** Make sure that the way you defend yourself fits the threat you are facing. Don't use too much force, and if you can, try to find answers without violence.
  - **Training and Responsibility:** You are responsible for making sure you know how to use defence tools correctly. Know what might happen and what the moral effects are of using force.

This part goes over all the different types of defence tools and methods, covering things like personal and home defence tools, useful defence strategies, and moral and legal issues. You can make your home safer and more ready for different kinds of emergencies by adding these steps to your home defence plan.

Creating a Safe Room: Planning and Execution

A safe room, sometimes called a panic room, is a reinforced room in your house that you can use in case of an emergency to feel safe and protected. For your safe room to meet your security needs, it's important that it's designed and built correctly.

1. Purpose and Objectives

- **Primary Functions:**
  - **Protection:** People who live in the safe room should be safe from burglars, natural disasters, and other situations.

- **Isolation:** It ought to provide a safe space where people can stay away from danger until help comes.
- **Communication:** To get in touch with family or emergency services, make sure the safe room has a way to communicate.

- **Assessment of Needs:**
  - **Threat Assessment:** Think about the kinds of threats you are most likely to face, like home invasions and bad weather. Make sure that the safe room has the right features to deal with these risks.
  - **Occupant Requirements:** Think about how many people will be using the safe room and what they might need (like medical goods or room for their pets).

2. Location and Design

- **Choosing a Location:**
  - **Central Location:** The safe room should be in the middle of your house, away from any outside walls or windows, so that it is less vulnerable to threats from outside.
  - **Accessibility:** Make sure that the spot is easy to get to from different rooms in the house. If water is a worry, stay away from basements.

- **Design Considerations:**
  - **Structural Reinforcement:** For the most safety, make sure the walls, doors, and ceilings are all reinforced. Think about steel, reinforced concrete, or ballistic-grade walls as possible materials.
  - **Door and Lock:** Put in a strong, heavy-duty door with a high-security lock. For extra safety, think about getting an electronic lock with a keypad or biometric entry.
  - **Ventilation:** Add a ventilation system or air filter to keep the air clean and keep people from suffocating or being exposed to dangerous situations.

3. Essential Features and Supplies

- **Emergency Supplies:**
  - **Food and Water:** Stock up on non-perishable food and enough water for everyone who will be living there. Try to get enough for at least 72 hours.
  - **First Aid Kit:** Include a full first aid kit with all the medical supplies, medicines, and other things that family members might need.
  - **Tools and Equipment:** Keep important tools like a fire extinguisher, a torch, batteries and a multi-tool in a safe place.

- **Communication Devices:**
  - **Phone or Radio:** Make sure the safe room has a reliable way to talk to people, like a home phone, a cell phone with a backup battery, or a two-way radio.
  - **Emergency Alerts:** Set up an emergency radio that runs on batteries or can be turned on and off by hand to get weather reports and emergency messages.
- **Comfort and Hygiene:**
  - **Seating:** Set up places to sit or sleep to make sure people are comfortable during long stays.
  - **Hygiene Supplies:** Bring things that are good for your health, like a portable toilet, cleaning supplies, and trash bags for getting rid of trash.

4. **Implementing Security Features**
   - **Reinforced Entry Points:**
     - **Walls and Ceiling:** Add extra layers of defence or structural improvements to the walls and ceilings to make them stronger.
     - **Window Protection:** If the safe room has windows, cover them with shatterproof or ballistic film to keep things from breaking and make the room safer.
   - **Escape Routes:**
     - **Secondary Exits:** Make sure the safe room has an extra exit or escape route in case the main entrance is broken into. Make sure the way out is hidden and easy to get to.
     - **Hidden Access:** For extra safety, you might want to add a hidden escape hatch or entry point. Make sure it's private and easy to use.
   - **Monitoring and Detection:**
     - **Security Cameras:** Put up security cams outside the safe room to keep an eye on the perimeter and spot any threats that might be coming.
     - **Motion Sensors:** Motion monitors should be used to let people inside the safe room know if there is any movement near the entrances.

5. **Training and Maintenance**
   - **Training:**
     - **Family Drills:** Do regular drills with your whole family to get them used to the safe room's location, features, and processes.

- **Emergency Procedures:** Go over and practise emergency plans, such as how to quickly get into the safe room and what to do once you're there.

- **Maintenance:**
  - **Regular Checks:** Check the safe room often to make sure that all of the supplies, tools, and safety features are still working.
  - **Restocking:** Regularly restock things that go bad and add new ones as needed. Check the dates to see if the things are still good, and if they are, replace them.

6. **Legal and Ethical Considerations**
   - **Local Building Codes:**
     - **Compliance:** Make sure that the building of the safe room follows all local building rules and codes. Get the appropriate permits if you need to.
     - **Professional Assistance:** To make sure the safe room meets all safety and security standards, you might want to talk to a professional builder or security expert.
   - **Ethical Use:**
     - **Non-violent Purpose:** You should only use the safe room for safety and protection. Do not use it to hold people against their will or for any other illegal reason.
     - **Accessibility:** Make sure that everyone in the family can get to the safe room, even those who have special needs or challenges.

This part gives you all the information you need to plan and set up a safe room in your home. By following these tips, you can make a safe and useful place that will give you peace of mind and protection in case of an emergency.

CHAPTER 6
INTERACTIONS IN TIMES OF CRISIS

Putting in Place Dependable Communication Mechanisms

Setting up reliable ways to communicate is important for staying in touch and working together during a problem. This part goes over the most important steps you need to take to set up and make sure your communication tools work.

1. **Assessing Your Communication Needs**

   - **Identify Requirements:**
     - **Family Size:** Find out how many people need to interact and what their specific needs are, like medical concerns or features that make it easier for them to use.
     - **Potential Scenarios:** Think about the different kinds of situations you could face, like natural disasters, power outages, or threats to your safety. Change the way you communicate to handle these situations.
   - **Communication Goals:**
     - **Reachability:** Make sure you can quickly get in touch with all of your family members and emergency contacts.
     - **Redundancy:** Use more than one way to communicate to avoid depending on a single point of failure.

2. **Establishing a Communication Hub**

   - **Central Location:**
     - **Designate a Spot:** Pick a central spot in your home that is easy to get to for your contact hub. This is where all of your contact tools and important documents should be kept.
     - **Organization:** Organise the space so that all the tools, chargers, and materials you need are easy to find.
   - **Essential Tools:**
     - **Cell Phones and Chargers:** In the contact hub, keep cell phones with extra batteries and chargers.

- **Emergency Radio:** Bring an emergency radio that can be powered by batteries or a hand crank so you can get messages and news.
- **Landline Phone:** As an alternative way to talk, keep a landline phone in the hub if you can.

3. Implementing Backup Power Solutions

- **Power Sources:**
  - **Generators:** During long power blackouts, use a generator to power important communication devices. Make sure the right steps are taken for setup, upkeep, and safety.
  - **Solar Chargers:** Solar-powered chargers can keep devices working even when the main power source isn't available.
- **Battery Management:**
  - **Stock Up:** Always have extra batteries on hand for important tools like telephones, flashlights, and more.
  - **Conservation:** To get more use out of your battery, do things like dimming screens and turning off features that aren't needed.

4. Setting Up Communication Devices

- **Cell Phones:**
  - **Backup Options:** Make sure you have extra batteries or chargers for your cell phones. Check these backup choices often to make sure they work.
  - **Emergency Apps:** Install and keep emergency apps that offer maps, alerts, and ways to communicate up to date.
- **Landline Phones:**
  - **Corded Phones:** As a backup, keep a corded landline phone handy, especially during power blackouts when cell networks might not work.
- **Two-Way Radios:**
  - **Selection:** Pick radios that have the right frequency and range for your needs. Make sure they can reach family members and emergency contacts within the area that was planned.
  - **Training:** Make sure everyone in the family knows how to use the radios and what the rules are.
- **Emergency Radio:**
  - **Features:** Choose an emergency radio that works with NOAA Weather Radio and has AM/FM channels to get full weather and emergency updates.

- **Power Options:** If you can, choose a radio that can be powered by batteries, a hand crank, or the sun.
- **Satellite Phones:**
  - **Usage:** When normal ways of communicating aren't available, you might want to think about getting a satellite phone. Make sure it's charged and works.

5. **Developing a Communication Plan**
   - **Emergency Contact List:**
     - **Create a List:** Make a list of important people to call, like family, neighbours, emergency services, and medical professionals.
     - **Update Regularly:** Update the contact list often to make sure it's correct and up-to-date.
   - **Check-In Procedures:**
     - **Routine Checks:** Set up regular times to check in with family, friends and emergency contacts. Make sure everyone knows when and how to log in.
     - **Emergency Signals:** Come up with and agree on emergency signs or codes that you can use to talk to each other when things are getting tough.

6. **Testing and Drilling**
   - **Communication Drills:**
     - **Regular Practice:** Hold regular drills to practise how to talk to people. Test how well your communication methods work by simulating different kinds of emergencies.
     - **Evaluate Performance:** Check how well your communication system worked during drills and make any changes that are needed based on what you learnt and what other people said.
   - **Device Checks:**
     - **Functionality Tests:** Check all communication gadgets on a regular basis to make sure they are still working. Check the chargers, batteries, and other important parts.

7. **Troubleshooting and Maintenance**
   - **Troubleshooting Issues:**
     - **Common Problems:** Be aware of common problems with communication, like devices that don't work or charges that die. Make a list of steps to take to fix these issues.
     - **Repair and Replacement:** Plan how you will fix or replace gadgets that don't work right. If you can, keep extra gadgets or spare parts on hand.

- **Ongoing Maintenance:**
  - **Regular Reviews:** Review and change your contact plan and setup on a regular basis. Make sure that all of the tools and methods you use are still useful and effective.
  - **Inventory Management:** Keep track of the communication tools and goods you have on hand and restock as needed.

8. **Legal and Safety Considerations**
   - **Regulations and Compliance:**
     - **Legal Requirements:** Make sure that all of your contact tools and methods are legal in your area.
     - **Safe Use:** When using communication tools, it's important to follow safety rules, especially when the power goes out or there are other dangerous situations.
   - **Privacy and Security:**
     - **Secure Communication:** When you can, use encrypted methods of contact to keep private data safe.
     - **Data Protection:** In a crisis, don't give out personal or private information without thinking first. Make sure that the ways you communicate are safe and private.

This part tells you everything you need to know to set up reliable contact systems in case of an emergency. You can make sure you stay connected, aware, and ready for emergencies by following these tips.

Digital security, radios, and emergency broadcasts

Having the right tools and knowing how to use them safely are both important for good communication during a crisis. This part talks about how to get reliable information from radios and emergency broadcasts, as well as how to keep your digital contact safe.

1. **Radios and Emergency Broadcasts**
   - **Types of Radios:**
   - ➤ **AM/FM Radios:**
     - **Usage:** You need an AM/FM radio to get neighbourhood news, weather reports, and emergency broadcasts. During a disaster, they can send information and alerts in real time.
     - **Power Options:** Choose radios that can be powered by batteries, hand crank, or solar power so that they will still work when the power goes out.
   - ➤ **NOAA Weather Radios:**

- **Purpose:** NOAA Weather Radios constantly send out weather reports from the National Weather Service, such as alerts and warnings for bad weather.
- **Features:** You should look for radios that can pick up NOAA Weather Radio and have emergency alert features.

> **Two-Way Radios:**

- **Function:** Walkie-talkies or two-way radios can be used for short-range contact, which is helpful when cell networks are down. You can stay in touch with family and neighbours through them.
- **Range and Channels:** Choose radios that have enough range for your needs and that have more than one channel to keep them from interfering with each other.

> **Satellite Radios:**

- **Capability:** Satellite radios can work in remote places where regular radio signals are weak or not available at all.
- **Subscription:** Keep in mind that most satellite radios need a subscription in order to fully view all of their channels and broadcasts.

- **Receiving Emergency Broadcasts:**

> **Emergency Alert System (EAS):**

- **Function:** The EAS sends out emergency messages and warnings on the radio, TV, and other channels. Make sure your radios can pick up these signals so you can get information quickly.
- **Frequency:** For EAS broadcasts and NOAA Weather Radio updates, make sure your radio is tuned to the right channels.

> **Local and National Alerts:**

- **Monitoring:** Keep an eye on local news and radio sources for emergency alerts on a regular basis. Listen to national news for more information about emergencies.
- **Emergency Broadcasts:** Listen to emergency reports for directions, safety tips, and orders to leave the area.

2. Digital Security

- **Protecting Communication Devices:**

> **Encryption:**

- **Secure Communication:** Utilise encrypted communication applications and devices to safeguard your messages and calls against unauthorised intrusion.

- **App Options:** It is advisable to utilise encrypted messaging applications that provide end-to-end encryption for secure communication in times of emergency.

➢ **Password Protection:**

- **Strong Passwords:** Utilise robust and distinctive passwords for your communication devices and applications. Refrain from utilising passwords that can be easily predicted or deduced.

- **Two-Factor Authentication (2FA):** Implement two-factor authentication (2FA) on your communication accounts to enhance the level of security.

- **Securing Digital Communication:**

➢ **Safe Practices:**

- **Avoid Public Wi-Fi:** It is advisable to refrain from utilising public Wi-Fi networks for transmitting sensitive information, as they may lack security measures and are susceptible to interception.

- **Use VPNs:** Utilise Virtual Private Networks (VPNs) to employ encryption on your internet connection and safeguard your online activities from surveillance.

➢ **Data Protection:**

- **Backup Data:** Ensure the regular preservation of crucial data, such as contact lists and conversation logs, by storing them securely.

- **Encryption:** Utilise encryption techniques to safeguard confidential files and communications, thereby thwarting unauthorised entry and guaranteeing confidentiality.

➢ **Handling Digital Threats:**

- **Phishing Scams:** Exercise vigilance against phishing scams and deceptive messages that aim to pilfer personal information. Prior to responding or clicking on links, it is important to authenticate the communications.

- **Software Updates:** Ensure that you regularly update your communication devices and applications with the most recent security patches and software upgrades in order to safeguard against potential vulnerabilities.

- **Emergency Digital Preparedness:**

➢ **Emergency Apps:**

- **Install Apps:** Download and install emergency applications that include notifications, navigational tools, and communication functionalities. Verify that they are up-to-date and operational.

- **App Settings:** Adjust the application's preferences to enable the reception of push notifications and alerts specifically for emergency situations.

- ➢ **Digital Literacy:**
  - **Education:** Acquire knowledge and instruct both yourself and your family on the subject of digital security practices and the significance of safeguarding personal information.
  - **Training:** Deliver instruction on the proper utilisation of secure communication applications and the identification and handling of digital hazards.

3. Combining Radio and Digital Security

- **Integrated Systems:**
  - **Use Together:** Combine radios and digital communication tools to enhance your information-gathering and communication capabilities.
  - **Backup Plans:** Make sure there are backup communication channels in case digital tools experience malfunctions or are compromised.
- **Redundancy:**
  - **Multiple Channels:** Employ many communication channels, including as radios, digital applications, and landlines, to guarantee that you can remain informed and linked in the event of a catastrophe.
- **Regular Updates:**
  - **Review Protocols:** Consistently evaluate and revise your communication and security measures to accommodate emerging threats and technological progress.

This section offers instructions on utilising radios for emergency broadcasts efficiently and ensuring digital security to safeguard your communication. By employing these strategies, you can guarantee the consistent transmission of accurate information and the safeguarding of communication during a crisis.

Keeping Updated Without Internet Access

During a crisis, the internet may become unattainable as a result of power blackouts, network malfunctions, or other disturbances. It is imperative to possess other means of staying informed and obtaining vital information. This section provides practical techniques and resources for staying informed in situations where internet access is not available.

1. Utilizing Battery-Powered and Hand-Crank Radios

- **Emergency Radio:**
  - **Features:** Acquire a battery-operated or hand-cranked emergency radio capable of receiving AM, FM, and NOAA Weather Radio transmissions. These radios are indispensable for obtaining emergency alerts and weather updates in the absence of internet connectivity.

- **Power Sources:** Make sure that the radio is equipped with various power alternatives, such as batteries, hand-crank, and solar power, in order to operate during prolonged periods of power loss.

- **Radio Frequencies:**
  - **NOAA Weather Radio:** Listen to NOAA Weather Radio broadcasts to receive weather warnings, emergency alerts, and updates from the National Weather Service.
  - **AM/FM Stations:** Tune in to nearby AM and FM radio stations to stay informed about current events, receive emergency alerts, and stay connected with your neighbourhood.

2. **Monitoring Emergency Broadcast Systems**
   - **Emergency Alert System (EAS):**
     - **Function:** The EAS transmits vital information, such as notifications and admonitions, via radio and television. Make sure that your radio or television is capable of receiving Emergency Alert System (EAS) transmissions in order to obtain timely updates.
     - **Reception:** Ensure that your devices are set to frequencies or channels that broadcast Emergency Alert System (EAS) notifications.
   - **Local News Channels:**
     - **Television:** Utilise a television that is powered by a battery or hand-crank mechanism to view local news broadcasts. This feature can deliver real-time updates regarding emergencies and offer directives issued by local authorities.
     - **Community Channels:** Keep watching community TV channels or public access stations for local updates and information.

3. **Using Traditional Media Sources**
   - **Printed Newspapers:**
     - **Subscription:** Ensure a continuous subscription to local newspapers or maintain a stock of recent publications. Newspapers offer vital insights into current events and the state of local affairs.
     - **Emergency Editions:** Look for special editions of newspapers that might be released during major emergencies.
   - **Bulletin Boards:**
     - **Community Boards:** Take note of the bulletin boards located in community centres, libraries, or local businesses. They might provide updates and notifications of local emergencies and available assistance.

4. **Leveraging Community Networks**

    - **Neighborhood Watch:**

        - **Coordination:** Participate in or create a neighbourhood watch or community group that collaborates to exchange information and stay informed during a crisis. These groups can offer vital information and assistance at the local level.

        - **Meetings:** Participate in community meetings or gatherings to remain updated on local circumstances and emergency protocols.

    - **Local Organizations:**

        - **Relief Agencies:** Seek updates and support by contacting local relief agencies, churches, or non-profit organisations. They can offer information regarding available resources, shelters, and emergency services.

        - **Community Contacts:** Establish a network of individuals in your local area who can provide you with relevant information and updates in the event of a disaster.

5. **Preparing and Using Physical Information Resources**

    - **Emergency Preparedness Guides:**

        - **Manuals:** Maintain hard copies of emergency preparedness guidelines, survival manuals, and local emergency protocols. These guides can provide essential information and directions in situations where other resources are not accessible.

        - **Checklists:** Utilise checklists for disaster readiness to guarantee that you possess all essential provisions and knowledge.

    - **Maps and Charts:**

        - **Local Maps:** Ensure the upkeep of tangible maps of your immediate vicinity, encompassing evacuation pathways, positions of emergency services, and locations of utility shut-off points. Maps are useful tools for navigating and planning in times of distress.

        - **Emergency Charts:** Maintain charts or diagrams that provide a clear and concise overview of emergency procedures, first aid protocols, and survival tactics.

6. **Maintaining Communication with Emergency Services**

    - **Emergency Contact List:**

        - **Phone Numbers:** Maintain a physical record of crucial emergency contacts, encompassing local emergency services, medical practitioners, and utility corporations. This list should be readily available and often updated.

- **Emergency Numbers:** Acquaint yourself with the emergency contact numbers that are specific to your locality, such as those for the police, fire department, and medical services.
- **Public Announcements:**
  - **Loudspeakers and Sirens:** Take heed of public announcements disseminated through loudspeakers or sirens. These may offer immediate information or directives from local authorities.

7. Training and Education

- **Emergency Drills:**
  - **Practice:** Regularly organise emergency exercises to simulate the experience of remaining informed and responding to situations without relying on the internet. Conduct simulations of various scenarios to assess your level of preparedness.
  - **Review:** Revise and enhance your emergency plans in accordance with the results of drills and any newly acquired knowledge.
- **Education:**
  - **Workshops:** Engage in community workshops or training sessions focused on emergency preparedness and acquiring survival skills. These can offer helpful information and resources.

8. Building Resilience and Resourcefulness

- **Self-Reliance:**
  - **Skills Development:** Acquire proficiency in navigation, first aid, and fundamental survival practices to augment your capacity to remain well-informed and secure without depending on digital resources.
  - **Adaptability:** Demonstrate flexibility and ingenuity in utilising accessible technologies and sources of information to remain well-informed in times of crisis.

This section provides pragmatic approaches for remaining updated without relying on the internet by utilising conventional media outlets, community networks, tangible resources, and emergency services. By employing and executing these strategies, you can guarantee that you stay up-to-date and knowledgeable in the event of any emergency scenario.

CHAPTER 7
HEALTH CARE READINESS AND BASIC FIRST AID

Keeping Essential Medicines in Stock

During a crisis, it is crucial to have the ability to obtain necessary medications in order to save one's health and effectively handle long-term health issues. Ensuring the proper stockpile of pharmaceuticals guarantees that you possess the essential resources to manage both regular and urgent health requirements. This section provides comprehensive guidance on the efficient accumulation of vital pharmaceuticals, encompassing factors such as medication types, storage methods, and overall management.

1. **Identifying Essential Medications**

   - **Chronic Condition Medications:**

     - **Personal Prescriptions:** Create a comprehensive compilation of pharmaceuticals used to treat chronic illnesses such as diabetes, hypertension, asthma, and heart disease. Encompass both prescribed and non-prescribed (OTC) drugs.

     - **Dosage and Frequency:** Document the prescribed dosage and frequency for each drug to enable efficient management of your pharmaceutical supply.

   - **Emergency Medications:**

     - **Antibiotics:** Keep a supply of versatile antibiotics that can effectively treat a wide range of common infections. Seek guidance from a healthcare professional to acquire prescriptions and gain a comprehensive understanding of correct usage.

     - **Pain Relievers:** Include analgesics such as ibuprofen and paracetamol for the purpose of managing pain and reducing fever.

     - **Anti-Allergy Medications:** If you or your family members suffer severe allergies, it is important to keep antihistamines and epinephrine auto-injectors readily available.

- **First Aid Supplies:**
  - **Topical Ointments:** Ensure the inclusion of antiseptic creams, burn ointments, and antibiotic ointments in order to effectively treat small wounds and skin disorders.
  - **Hydrocortisone Cream:** Hydrocortisone cream is available for the purpose of alleviating itchiness and irritation caused by skin rashes or insect bites.

2. **Procuring Medications**
   - **Consulting Healthcare Providers:**
     - **Prescriptions:** Acquire prescriptions from your healthcare physician for vital medications, particularly those necessary for long-term illnesses or urgent circumstances.
     - **Medication Review:** Engage in a conversation with your healthcare physician regarding any supplementary medications you may require and the process of acquiring them.
   - **Pharmacy Management:**
     - **Refills:** Establish automated replenishment for essential prescriptions that are vital to your well-being. Enquire with your pharmacy on the availability of extended refill alternatives.
     - **Over-the-Counter Medications:** Consider buying over-the-counter (OTC) medications in large quantities, particularly for prevalent conditions such as colds, headaches, and stomach problems.
   - **Online and Emergency Resources:**
     - **Online Pharmacies:** Consider utilising internet pharmacies as an alternative means of acquiring pharmaceuticals in situations where local options are scarce. Ensure that they have a good reputation and adhere to safety regulations.
     - **Emergency Kits:** Ensure that your emergency preparedness kit contains an ample quantity of vital drugs.

3. **Proper Storage of Medications**
   - **Storage Conditions:**
     - **Temperature Control:** Ensure that drugs are stored within the prescribed temperature range. Several drugs require storage in a cold and dry location, shielded from direct sunlight.
     - **Humidity Control:** Utilise hermetically sealed containers to safeguard pharmaceuticals against humidity and moisture, as these factors can compromise their efficacy.
   - **Shelving and Containers:**
     - **Organized Storage:** Utilise transparent, properly identified receptacles to categorise pharmaceuticals based on their type and expiration date. Store them in a designated area that is conveniently reachable.

- **Childproofing:** Make sure to put all drugs in containers that are designed to be difficult for children to open, and keep them in a location where children cannot access.

- **Expiration Dates:**
  - **Monitoring:** Frequently inspect the expiration dates of all medications and prioritise the use of older prescriptions by rotating the supply. Properly discard expired or unused pharmaceuticals in accordance with local legislation.

4. Managing Medication Supply

- **Inventory Management:**
  - **Record Keeping:** Ensure that you keep a record of your medications, including the number of doses available, the dates of expiration, and the timetables for refilling them. Utilise a spreadsheet or a tangible diary to meticulously monitor your inventory.
  - **Regular Review:** Regularly assess your drug inventory to verify that you possess sufficient quantities and promptly resolve any deficiencies.

- **Emergency Scenarios:**
  - **Planning:** Develop contingency plans to address situations when there may be restricted access to pharmaceuticals, such as in the event of prolonged power outages or problems in transportation.
  - **Alternate Sources:** Identify prospective avenues for acquiring emergency medications, such as local health agencies, community organisations, or online resources.

- **Medication Sharing:**
  - **Legal and Safety Considerations:** Keep in mind that sharing prescription drugs is against the law. Sharing medications should only be done as told by a doctor and as allowed by law.

5. Special Considerations

- **Pediatric and Geriatric Medications:**
  - **Children's Medications:** Having paediatric versions of important medicines on hand is important if you have kids. Make sure you have the right dosages and guidelines for how to give them.
  - **Geriatric Needs:** Think about what your elderly family members need, like medicines to treat conditions that come with getting older and making sure the right dosage changes are made.

- **Mental Health Medications:**
  - **Psychiatric Medications:** If it applies, list any medicines you take for mental health problems, like antidepressants or anxiety pills. Make sure you have enough tools and a good place to store them.

- **Medical Devices and Supplies:**
  - **Supportive Devices:** Ensure an adequate inventory of crucial medical devices used for drug administration, including inhalers, insulin pens, and blood glucose monitors.
  - **Supplies:** Ensure to include additional medical supplies, such as syringes or test strips, that are necessary for the administration or monitoring of drugs.

This section offers extensive assistance on the process of stockpiling vital pharmaceuticals. It covers several aspects such as identifying the prescriptions that are necessary, acquiring and storing them, effectively managing the supplies, and addressing any particular considerations that may arise. By according to these instructions, you may guarantee that you are adequately equipped to manage medical requirements in the event of an emergency.

Putting Together a Complete First Aid Kit

A well-stocked first aid kit is essential for managing injuries and health emergencies during a crisis. It should be tailored to meet the needs of your household and be easily accessible. This section covers the critical components of a comprehensive first aid kit, including basic supplies, advanced tools, and how to assemble and maintain it.

1. Basic First Aid Supplies

- **Adhesive Bandages:**
  - **Variety of Sizes:** Include various sizes of adhesive bandages for minor cuts and scrapes. Ensure you have a mix of small, medium, and large bandages.
  - **Specialty Bandages:** Add waterproof bandages and blister pads for specific needs.
- **Sterile Gauze Pads and Adhesive Tape:**
  - **Gauze Pads:** Include sterile gauze pads in different sizes for wound care and dressing.
  - **Adhesive Tape:** Use medical adhesive tape to secure gauze pads and bandages in place.
- **Antiseptic Wipes and Ointments:**
  - **Antiseptic Wipes:** Pack antiseptic wipes or alcohol pads for cleaning wounds and preventing infection.
  - **Antibiotic Ointments:** Include antibiotic ointments to apply to minor cuts and abrasions.
- **Scissors and Tweezers:**
  - **Scissors:** Add medical scissors for cutting tape, gauze, and bandages.
  - **Tweezers:** Include tweezers for removing splinters and debris from wounds.
- **Thermometer:**
  - **Digital Thermometer:** Use a digital thermometer for accurately measuring body temperature.

- **Disposable Gloves:**
  - **Latex or Nitrile Gloves:** Pack disposable gloves to maintain hygiene while treating wounds and performing first aid.

2. **Advanced Medical Supplies**
   - **Splints:**
     - **Finger Splints:** Include finger splints for immobilizing injured fingers.
     - **Universal Splints:** Add flexible, adjustable splints for stabilizing broken limbs or severe injuries.
   - **Burn Dressings:**
     - **Burn Gel:** Pack burn gel or dressings specifically designed to treat burns and reduce pain.
   - **Eye Wash and Eye Pads:**
     - **Eye Wash:** Include sterile saline eye wash for rinsing out foreign objects or chemicals from the eyes.
     - **Eye Pads:** Add eye pads for covering and protecting injured or irritated eyes.
   - **Emergency Blanket:**
     - **Thermal Blanket:** Pack a thermal or Mylar blanket to provide warmth and prevent hypothermia.

3. **Medications and Supplements**
   - **Pain Relievers:**
     - **Ibuprofen and Acetaminophen:** Include over-the-counter pain relievers for managing pain and fever.
   - **Anti-Allergy Medications:**
     - **Antihistamines:** Pack antihistamines for treating allergic reactions and symptoms.
   - **Anti-Diarrheal and Laxatives:**
     - **Anti-Diarrheal Medications:** Include medications for managing diarrhea, such as loperamide.
     - **Laxatives:** Add laxatives for constipation relief.
   - **Oral Rehydration Salts:**
     - **Electrolyte Powders:** Include oral rehydration salts or electrolyte powders to prevent dehydration during illness or heat exposure.

4. **Specialized Medical Tools**
    - **Stethoscope:**
        - **Basic Stethoscope:** Use a basic stethoscope to listen to heartbeats and lung sounds for basic health assessments.
    - **Blood Pressure Cuff:**
        - **Manual Blood Pressure Monitor:** Include a manual blood pressure cuff for monitoring blood pressure.
    - **Surgical Tape and Sterile Dressings:**
        - **Surgical Tape:** Pack surgical tape for securing dressings and bandages.
        - **Sterile Dressings:** Include larger sterile dressings for covering larger wounds or injuries.

5. **Creating and Maintaining Your Kit**
    - **Kit Container:**
        - **Durable and Portable Container:** Use a durable, waterproof container to keep your first aid supplies organized and protected. A large, clear plastic box or a specialized first aid bag works well.
    - **Labeling and Organization:**
        - **Labeling:** Clearly label compartments or containers within your kit for easy identification of supplies.
        - **Organization:** Organize items by type (e.g., wound care, medications, tools) to ensure quick access in emergencies.
    - **Regular Inventory Check:**
        - **Expiration Dates:** Regularly check the expiration dates of medications and replace expired items.
        - **Restocking:** Replenish used supplies and add new items as needed. Perform a routine check every few months to ensure the kit is fully stocked.
    - **Training and Practice:**
        - **First Aid Training:** Take a first aid and CPR course to learn how to use the items in your kit effectively.
        - **Drills:** Conduct practice drills to familiarize yourself with the kit's contents and ensure everyone in the household knows how to use it.

6. **Customizing Your Kit**
    - **Family-Specific Needs:**
        - **Pediatric Supplies:** Include items for children, such as pediatric medications and smaller bandages.

- **Geriatric Supplies:** Add supplies tailored to elderly family members, such as medications for age-related conditions.
- **Special Medical Needs:**
  - **Chronic Conditions:** Stock additional supplies for chronic conditions specific to household members, such as insulin for diabetes or inhalers for asthma.
  - **Medical Devices:** Include any necessary medical devices or supplies, such as blood glucose monitors or EpiPens.
- **Emergency Scenarios:**
  - **Environmental Hazards:** Consider including items for specific environmental hazards, such as insect sting kits for outdoor emergencies or water purification tablets.

This section provides comprehensive guidance on assembling a first aid kit, including essential supplies, advanced medical tools, and tips for creating and maintaining your kit. By ensuring your first aid kit is well-stocked and organized, you can be better prepared to manage health emergencies during a crisis.

Medical Knowledge That Everyone Should Have

During a crisis, possessing fundamental medical expertise is essential for effectively handling injuries and health concerns in the absence of professional medical assistance. This section provides comprehensive instruction on fundamental medical skills that are crucial for properly managing typical emergency situations and enhancing overall readiness.

1. Basic First Aid Techniques

- **Wound Care:**
  - **Cleaning Wounds:** Acquire knowledge on the correct method of cleansing wounds by utilizing sterile water or saline to eliminate foreign particles and minimize the likelihood of infection.
  - **Applying Dressings:** Learn the proper technique for using sterile gauze or bandages to manage bleeding and safeguard wounds.
  - **Managing Infection:** Identify indications of infection, such as redness, swelling, and pus, and be aware of when it is necessary to seek medical assistance.
- **Burn Treatment:**
  - **Cooling Burns:** Utilize cool water, rather than cold water, to alleviate burns and mitigate pain. Refrain from utilizing ice or really cold water, as it has the potential to exacerbate the harm.

- **Applying Burn Dressings:** Administer aseptic burn dressings or burn gel to alleviate and safeguard the impacted region. It is imperative to promptly seek medical assistance for severe burns.

- **Splinting:**

    - **Immobilizing Injuries:** Acquire the knowledge of utilizing splints or makeshift materials to render fractured bones or sprains immobile, so alleviating discomfort and averting additional harm.

    - **Securing Splints:** Utilize bandages or adhesive tape to firmly fasten splints, maintaining their stability and preventing any further discomfort.

2. Cardiopulmonary Resuscitation (CPR)

- **Performing CPR:**

    - **Adult CPR:** Find out how to do CPR on an adult, such as how to do chest compressions and rescue breathing. At least 100 to 120 times per minute, press down on the chest until it's at least 2 inches deep.

    - **Child and Infant CPR:** Comprehend the disparities in CPR methodologies for children and infants, encompassing the utilization of a reduced number of chest compressions and more delicate rescue breaths.

- **Using an Automated External Defibrillator (AED):**

    - **AED Operation:** Acquaint yourself with the proper usage of an AED, which involves correctly positioning electrode pads and adhering to the device's instructions to provide a shock if necessary.

3. Basic Life Support (BLS)

- **Recognizing Emergencies:**

    - **Identifying Symptoms:** Acquire the ability to identify indications of medical crises, such as myocardial infarctions, cerebrovascular accidents, and hypersensitivity responses.

    - **Activating Emergency Services:** Acquire knowledge about how to reach emergency services and effectively communicate accurate and succinct details about the incident.

- **Managing Shock:**

    - **Recognizing Shock:** Recognize signs of shock, such as pallor, tachycardia, and cognitive disorientation. Maintain a state of tranquillity and ensure the individual's physical well-being.

    - **Treating Shock:** Position the individual in a horizontal position, if feasible, raise their legs, and use a blanket to preserve their body heat.

4. **Wound and Injury Management**
   - **Bleeding Control:**
     - **Applying Pressure:** Apply direct pressure to bleeding wounds using sterile cloths or bandages. If feasible, elevate the affected region to minimize blood circulation.
     - **Tourniquets:** Acquire the knowledge and skills necessary to properly utilize a tourniquet in cases of severe limb injuries when direct pressure is inadequate. Make careful to apply it directly on the cut and with enough pressure to halt the bleeding.
   - **Fracture and Dislocation Care:**
     - **Recognizing Fractures:** Comprehend the indications of fractures and dislocations, such as physical distortion, inflammation, and intense agony.
     - **Immobilization:** Immobilize the injured area using splints or sturdy objects. Avoid moving the injured person unless necessary.

5. **Identifying and Managing Common Health Concerns**
   - **Allergic Reactions:**
     - **Symptoms:** Identify symptoms of allergic reactions, such as urticaria, edema, and respiratory distress.
     - **Epinephrine Use:** Acquire the knowledge on how to utilize an epinephrine auto-injector to address severe allergic responses, specifically anaphylaxis. Adhere to the guidelines provided on the device.
   - **Heat and Cold Injuries:**
     - **Heat Exhaustion and Heatstroke:** Recognize signs such as profuse perspiration, nausea, and cognitive disorientation. Relocate to a more temperate environment, replenish fluids, and administer cold compresses.
     - **Frostbite and Hypothermia:** Identify symptoms such as sensory loss and tremors. Gradually apply heat to the affected areas and consult a medical professional if needed.
   - **Choking Relief:**
     - **Heimlich Maneuver:** Acquire the knowledge of the Heimlich maneuver, a technique used to alleviate choking in both adults and children. Administer abdominal thrusts to forcibly remove the obstruction obstructing the respiratory passage.
     - **Back Blows and Chest Thrusts:** To dislodge the object in newborns, employ a mix of back strikes and chest thrusts.

6. **Administering Medications and Supplements**
   - **Dosage and Administration:**
     - **Correct Dosage:** Acquire the knowledge and skills to accurately measure and administer drugs by according to dosage guidelines and utilizing suitable measuring instruments.
     - **Medication Interactions:** Exercise caution with potential drug interactions and refrain from combining medications without professional assistance.
   - **Using Medical Devices:**
     - **Inhalers and Insulin Pens:** Acquire the knowledge on the correct utilization of inhalers for asthma and insulin pens for diabetes. Adhere to the guidelines provided for the proper dosage and administration.

7. **Preparing for Specific Medical Needs**
   - **Chronic Conditions:**
     - **Managing Conditions:** Comprehend the fundamental principles of managing chronic illnesses such as diabetes, hypertension, and asthma. Ensure that essential medications and equipment are easily accessible.
   - **Pediatric Care:**
     - **Children's Health:** Acquire specialized methods for addressing injuries and illnesses in children, which encompass modifying dosages and implementing steps to enhance comfort.
   - **Geriatric Care:**
     - **Elderly Health:** Take into account the distinct healthcare requirements of older family members, which encompass challenges related to physical movement, persistent ailments, and the organization of medications.

8. **Continuing Education and Training**
   - **First Aid and CPR Courses:**
     - **Certification:** Register for first aid and CPR classes provided by reputable organizations such as the Red Cross or local community centres. Acquire certification to showcase your proficiency.
   - **Regular Practice:**
     - **Skill Refreshment:** Regularly evaluate and rehearse first aid methods to uphold skillfulness and assurance in critical circumstances.

- **Staying Informed:**
  - **Updates:** Stay informed about the most recent protocols and suggestions for initial medical assistance and urgent treatment by referring to trustworthy sources and respected medical institutions.

This part offers a thorough exposition of fundamental medical competencies that are important for all individuals to possess, encompassing rudimentary emergency care, cardiopulmonary resuscitation (CPR), advanced life support techniques, and specialized treatment for diverse health conditions. By acquiring and honing these abilities, you can enhance your readiness to handle medical emergencies and successfully assist your family in times of disaster.

CHAPTER 8
SUSTAINING HYGIENE AND SANITATION

Solutions for Waste Management for Extended Bug-Ins

Efficient waste management is crucial for preserving cleanliness and well-being in a prolonged bug-in scenario. Effective waste management mitigates pollution, minimizes potential health hazards, and promotes the establishment of a sustainable habitat. This section offers tactics and remedies for effectively handling different categories of waste over a prolonged duration.

1. **Types of Waste and Their Management**

   - **Household Waste:**
     - **Types:** Comprises of organic waste, wrapping materials, non-reusable objects, and overall domestic garbage.
     - **Management:** Categorize home garbage into recyclables, compostables, and non-recyclables, then handle each category in accordance with optimal methods.

   - **Organic Waste:**
     - **Types:** Organic trash, including food leftovers, yard garbage, and biodegradable materials.
     - **Management:** Utilize the process of composting to transform organic waste into compost that is abundant in nutrients for gardening purposes, while simultaneously decreasing the amount of garbage that requires disposal.

   - **Human Waste:**
     - **Types:** Excreted waste products in the form of urine and feces.
     - **Management:** Utilize suitable sanitation methods to effectively and hygienically manage human waste.

   - **Hazardous Waste:**
     - **Types:** Batteries, chemicals, and other toxic materials.

- **Management:** Ensure proper containment of hazardous waste and adhere to local regulations for its safe disposal.

2. Waste Management Techniques

- **Composting:**
  - **Setup:** Construct a composting system by utilizing either a compost bin or pile. Combine a combination of green (containing nitrogen) and brown (containing carbon) substances to the compost.
  - **Maintenance:** Regularly agitate the compost to introduce air and accelerate the process of decomposition. Maintain the moisture level of the compost at an appropriate level, avoiding excessive wetness.

- **Recycling:**
  - **Sorting:** Sort recyclable materials such as paper, cardboard, glass, and specific types of plastics into different categories. Refer to the local recycling guidelines for precise specifications.
  - **Storage:** Organize and store recyclables in designated containers or bags so that they are prepared and ready for recycling when facilities become accessible.

- **Waste Disposal:**
  - **Trash Bags:** Utilize long-lasting, impermeable garbage bags for rubbish that cannot be recycled. Ensure a secure seal on bags to effectively prevent the escape of odors and the intrusion of pests.
  - **Waste Collection:** Consistently gather and oversee rubbish to avoid buildup. If regular garbage disposal services are not accessible, it is advisable to explore alternate methods such as incineration (while adhering to safety and environmental protocols) or burial of non-recyclable waste.

3. Human Waste Management Solutions

- **Portable Toilets:**
  - **Types:** Utilize portable toilets or camping toilets that are equipped with waste disposal bags or chemicals. Select solutions that are user-friendly and require minimal upkeep.
  - **Maintenance:** Regularly empty and clean the portable toilet. Follow manufacturer instructions for safe disposal of waste.

- **Composting Toilets:**
  - **Operation:** Install a composting toilet that converts human waste into compost. Ensure adequate air circulation and utilize composting additives if necessary.
  - **Maintenance:** Adhere to the system's rules while handling composting toilet waste. Monitor the conditions of composting to guarantee efficient breakdown.

- **Emergency Solutions:**
  - **Bucket Toilets:** Utilize a container equipped with a cover and a protective lining for unforeseen circumstances. Place a disposable or biodegradable liner inside the bucket and include absorbent materials to effectively handle trash.
  - **Disposal:** Properly discard waste by burying it at a depth of at least 6-8 inches in a designated area that is located away from any water sources. Ensure that the burial place is not susceptible to inundation.

4. **Dealing with Hazardous Waste**
   - **Storage:**
     - **Containers:** Utilize impermeable, clearly marked receptacles for the storage of dangerous substances. Store containers in a safe and properly ventilated location.
     - **Safety:** Ensure that hazardous products are kept out of reach of youngsters and pets. Adhere to safety protocols to avoid any instances of leaks or spills.
   - **Disposal:**
     - **Guidelines:** Adhere to the specific rules and guidelines set by your local authorities when it comes to getting rid of dangerous waste materials. If there are no disposal services available, it is advisable to seek information from local authorities regarding the appropriate methods of processing and disposing of the waste.

5. **Waste Management Planning**
   - **Inventory and Supplies:**
     - **Stockpile:** Maintain a record of waste management resources, such as composting materials, recycling containers, garbage bags, and sanitation items.
     - **Rotation:** Periodically inspect and rotate provisions to guarantee their efficacy and discard any outdated or depleted products.
   - **Family Involvement:**
     - **Education:** Instruct family members on appropriate garbage disposal methods and obligations. Ensure universal comprehension of the proper segregation and management of various categories of garbage.
     - **Roles:** Allocate specific roles for waste management tasks to divide duties and uphold efficiency.
   - **Emergency Procedures:**
     - **Backup Plans:** Create contingency strategies for waste management in the event of supply shortages or system failures. Explore other approaches for garbage disposal if necessary.

This section provides comprehensive strategies for managing waste during a long-term bug-in situation, including handling household waste, organic waste, human waste, and hazardous materials. By implementing these waste management solutions, you can maintain a clean, healthy, and sustainable living environment.

Maintaining Personal Hygiene: Keeping Clean and Well

Ensuring personal cleanliness is crucial for maintaining good health and well-being, particularly in a prolonged bug-in scenario where resources may be scarce. Adhering to proper hygiene practices can effectively prevent disease, minimize the likelihood of infections, and enhance overall comfort. This section offers practical guidance on upholding personal hygiene in situations when traditional resources and facilities may not be accessible.

1. **Importance of Personal Hygiene**

   - **Health Benefits:**

     - **Disease Prevention:** Adhering to regular hygiene habits is crucial in preventing the transmission of infectious diseases and minimizing the likelihood of contracting illnesses caused by bacteria, viruses, and parasites.

     - **Comfort:** Maintaining cleanliness and proper grooming enhances physical comfort and emotional well-being, particularly in times of stress.

   - **Social Considerations:**

     - **Community Well-Being:** Practicing good personal hygiene promotes a healthier living environment and facilitates better relationships within a family or community setting.

2. **Hand Hygiene**

   - **Hand Washing:**

     - **Technique:** Thoroughly cleanse your hands using soap and water for a minimum of 20 seconds, ensuring that you cover all surfaces, including the front, back, spaces between fingers, and underneath nails.

     - **Frequency:** Prior to consuming food, after utilizing the restroom, and following contact with waste or soiled objects, it is vital to cleanse one's hands.

     - **Alternative Methods:** In the absence of soap and water, utilize alcohol-based hand sanitizers with a minimum of 60% alcohol. Ensure an ample amount of sanitizer is applied to completely cover all parts of the hands, and vigorously rub until the sanitizer has evaporated.

   - **Hand Care:**

     - **Moisturizing:** Frequent hand washing might cause skin dehydration. Apply hand lotion or moisturizer regularly to maintain proper hydration of the hands and avoid the occurrence of cracks and discomfort.

- **Nail Care:** Maintain well-groomed and hygienic nails to avoid the accumulation of dirt and bacteria. Refrain from gnawing on your nails or plucking at your cuticles.

3. Body Hygiene

- **Bathing and Showering:**
  - **Frequency:** Regularly cleanse your body by bathing or showering to eliminate sweat, grime, and bacteria. If there is a scarcity of water, it is advisable to take sponge baths or use wipes to cleanse essential areas.
  - **Water Conservation:** Conserve water by minimizing the duration of your showers or baths. If needed, gather and recycle water for various hygienic activities.

- **Cleaning Products:**
  - **Soap and Shampoo:** Utilize mild soap and shampoo to purify the skin and hair. If traditional products are not accessible, utilize natural substitutes such as baking soda or vinegar.
  - **DIY Solutions:** Produce DIY washing solutions using organic components, such as a blend of water, vinegar, and a little quantity of essential oil for a body wash.

4. Oral Hygiene

- **Brushing and Flossing:**
  - **Technique:** It is recommended to brush your teeth at least twice a day using toothpaste that contains fluoride. Utilize a toothbrush with soft bristles and engage in brushing for a minimum duration of two minutes. Engage in daily flossing to eliminate plaque and food debris that accumulate between the teeth.
  - **Alternative Products:** In the absence of toothpaste, one can utilize a combination of baking soda and water as a temporary alternative. If a toothbrush is unavailable, utilize clean and soft rags or sticks as an alternative.

- **Dental Care:**
  - **Rinsing:** To uphold dental hygiene and alleviate gum discomfort, employ a saltwater rinse or a self-made mouthwash, such as a blend of water, baking soda, and salt.
  - **Dentures and Bridges:** Follow the directions on the box to clean your dentures or other oral appliances. After each meal, wash them well and put them away in a clean, dry place.

5. **Hair and Skin Care**

    - **Hair Care:**

        - **Washing:** To maintain cleanliness and manageability, it is advisable to routinely cleanse the hair using shampoo or a comparable substitute. If there is a scarcity of water, you may want to contemplate using dry shampoo or powder.

        - **Combing and Brushing:** Utilize a comb or brush to disentangle hair and eliminate any foreign particles. Consistent grooming is essential for preserving the health of the hair and minimizing scalp irritation.

    - **Skin Care:**

        - **Cleansing:** Utilize gentle soap or organic cleansers to cleanse the skin and eliminate dirt and sebum. Give particular focus to areas that are susceptible to perspiration and bacterial growth.

        - **Moisturizing:** To avoid dry skin, particularly in challenging environments or when water is scarce, it is advisable to use lotion or moisturizer. Utilize organic oils, such as coconut oil, as an alternative when commercial products are inaccessible.

6. **Clothing and Laundry**

    - **Clean Clothing:**

        - **Frequency:** Regularly change your clothes to uphold cleanliness and avoid the accumulation of perspiration and microorganisms. It is advisable to wear fresh underwear and socks on a daily basis.

        - **Layering:** Utilize the technique of layering garments to effectively regulate body temperature, taking into account variations in physical exertion and surrounding environmental factors.

    - **Laundry Practices:**

        - **Washing:** Cleanse garments, linens, and towels as necessary. If a washing machine is accessible, utilize it; otherwise, manually wash the items using soap and water.

        - **Drying:** Ensure clothing are completely dried to prevent the growth of mold and mildew. If a dryer is unavailable, air dry the goods by using indoor drying racks or hanging the garments in a well-ventilated place.

7. **Personal Hygiene Supplies**

    - **Stocking Up:**

        - **Essential Items:** Maintain a stock of personal hygiene items, including soap, shampoo, toothpaste, dental floss, lotion, and feminine hygiene goods.

        - **Storage:** To preserve the efficacy of hygiene supplies and guarantee their accessibility, it is important to store them in a pristine and moisture-free location.

- **DIY and Alternatives:**

  - **Homemade Products:** Make your own personal care products with readily available household components. For instance, vinegar can be utilized as a hair rinse, while baking soda can be employed for tooth cleaning.

  - **Conservation:** Utilize hygiene items in moderation and investigate alternative approaches to uphold cleanliness in order to prolong the accessibility of vital resources.

8. Hygiene in Shared Spaces

- **Shared Facilities:**

  - **Cleaning:** Consistently sanitize communal places such as restrooms, kitchens, and common living spaces to uphold cleanliness and avert the spread of germs amongst individuals.

  - **Protocols:** Implement and adhere to hygiene rules for communal areas, which involve practicing hand hygiene by washing hands before and after use shared facilities.

- **Health Monitoring:**

  - **Observations:** Supervise the well-being of all family members for indications of sickness or contagion. Take immediate action to address any hygiene concerns in order to prevent the transmission of illnesses.

This section provides detailed instructions on how to maintain personal hygiene in a prolonged bug-in scenario. It covers several aspects such as hand cleaning, body care, oral hygiene, and the management of clothing and laundry. Adopting these habits aids in guaranteeing cleanliness, well-being, and comfort in difficult situations.

Avoiding Illness in Tight Spaces

The proximity of individuals in a prolonged bug-in scenario heightens the likelihood of disease transmission. Implementing effective preventative methods is crucial for preserving health and minimizing the transmission of illnesses within families or groups. This section presents pragmatic strategies to mitigate the spread of diseases in enclosed living environments.

1. Understanding Disease Transmission

- **Modes of Transmission:**

  - **Airborne:** Respiratory illnesses, including colds, influenza, and COVID-19, are transmitted through airborne droplets.

  - **Contact:** Pathogens are transmitted by direct contact with infected surfaces or bodily fluids, encompassing gastrointestinal illnesses and dermatological diseases.

- **Vector-Borne:** Vector-borne infections, such as Lyme disease or hantavirus, are transmitted by insects or animals, such as ticks, mosquitoes, or rats.
- **Risk Factors:**
  - **Crowding:** Greater population density heightens the probability of disease transmission.
  - **Shared Facilities:** Shared use of restrooms, kitchens, and other spaces can promote the spread of disease.

2. Hygiene Practices

- **Hand Hygiene:**
  - **Regular Washing:** Regularly cleanse hands using soap and water, particularly prior to consuming food, after utilizing the restroom, and following coughing or sneezing.
  - **Hand Sanitizer:** Utilize alcohol-based hand sanitizers in situations when soap and water are not accessible. Make sure to completely cover your hands and let the sanitizer dry.
- **Surface Cleaning:**
  - **Disinfecting:** Frequently sanitize and disinfect frequently touched surfaces such as doorknobs, light switches, and counters using disinfectants that are capable of eliminating germs and viruses.
  - **Cleaning Schedule:** Implement a regular cleaning regimen and delegate specific responsibilities to ensure constant cleanliness in all locations.
- **Personal Hygiene:**
  - **Bathing:** Ensure personal hygiene by engaging in regular bathing or utilizing sponge baths in situations when water availability is limited.
  - **Grooming:** Maintain well-groomed nails and refrain from biting or plucking them to minimize the likelihood of infections.

3. Managing Respiratory Health

- **Avoiding Germ Spread:**
  - **Covering Coughs and Sneezes:** Utilize tissues or the crook of the elbow to shield coughs and sneezes. Promptly discard used tissues and thereafter cleanse hands.
  - **Face Masks:** It is advisable to utilize face masks as a means to minimize the transmission of airborne diseases, particularly when an individual within the home is unwell.
- **Ventilation:**
  - **Air Circulation:** Enhance the quality of indoor air by implementing proper ventilation in enclosed areas. Maximize ventilation by opening windows whenever feasible and utilizing fans to enhance air circulation.

- **Air Purifiers:** Utilize air purifiers equipped with HEPA filters to diminish the presence of airborne infections and allergens.

4. Food and Water Safety

- **Food Handling:**
  - **Cleanliness:** It is important to cleanse your hands both before and after coming into contact with food. Ensure that you utilize hygienic tools and surfaces when preparing food.
  - **Storage:** Ensure proper storage of perishable items to avoid spoiling and infection. Adhere to the instructions for refrigeration and utilize airtight containers.

- **Water Safety:**
  - **Clean Water:** Ensure that water sources are pristine and devoid of any form of contamination. Utilize water filtration systems as needed and consistently inspect for any indications of pollution.
  - **Safe Practices:** Refrain from utilizing water that has been tainted for the purposes of cooking or consuming. Ensure that water is stored in sterile and airtight containers to avoid any form of pollution.

5. Managing Waste

- **Proper Disposal:**
  - **Segregation:** Categorize garbage into distinct groups, including recyclable materials, compostable items, and non-recyclable substances. Utilize suitable disposal techniques for each respective category.
  - **Handling:** Promptly discard rubbish to prevent its accumulation and minimize the likelihood of attracting bugs.

- **Sanitation:**
  - **Cleaning Up:** Frequently cleanse and disinfect waste receptacles and locations where garbage is stored or handled.
  - **Pest Control:** Enforce strategies to mitigate the presence of pests, such as ensuring proper food containment and employing traps or repellents.

6. Disease Monitoring and Response

- **Health Monitoring:**
  - **Symptoms:** Observe for signs of illness, such as elevated body temperature, respiratory distress, or digestive problems. Make a record of any atypical symptoms or alterations in your health.
  - **Health Records:** Maintain a comprehensive health record for each member of the family, documenting any indications of illness, remedies administered, and medical background.

- **Medical Care:**
  - **First Aid:** Ensure that you are ready to administer initial medical assistance for small wounds or ailments. Acquire knowledge about fundamental treatment techniques and understand when it is necessary to seek assistance from a qualified medical practitioner.
  - **Emergency Resources:** Ensure that you have a strategy in place to obtain medical assistance if necessary. Locate and determine the availability of nearby health facilities and develop effective means of communication for emergency situations.

7. **Educating and Preparing**
   - **Training:**
     - **Hygiene Education:** Instruct all members of the home about appropriate hygiene protocols and the significance of disease prevention.
     - **Emergency Procedures:** Formulate and implement emergency protocols for effectively handling outbreaks of illness or preventing the spread of diseases.
   - **Preparation:**
     - **Supplies:** Ensure you have an ample supply of necessary hygiene and medical products, such as hand sanitizers, disinfectants, first aid kits, and prescriptions.
     - **Information:** Keep yourself updated regarding any illness outbreaks and health advisories. Modify preventive tactics in accordance with up-to-date data.

This section offers extensive tactics for illness prevention in confined spaces, encompassing practices such as upholding hygiene, regulating respiratory well-being, guaranteeing the safety of food and drink, and monitoring overall health. Enforcing these procedures aids in protecting health and mitigating the transmission of diseases in limited living environments.

CHAPTER 9
CRUCIAL SKILLS FOR SURVIVAL

Starting a Fire and Preserving Heat

During a survival scenario, it is imperative to possess the skill of igniting a fire and conserving heat as it is essential for providing warmth, preparing food, and upholding morale. This section provides crucial strategies for igniting fires and approaches for maximizing heat retention.

1. **Fire-Starting Techniques**

   - **Traditional Methods:**

     - **Matches and Lighters:** The most direct and dependable approaches. To maintain the functionality of matches and lighters under unfavourable conditions, it is advisable to store them in waterproof containers.

     - **Fire Starters:** Utilize commercial fire igniters such as magnesium blocks or fire sticks or create DIY igniters by saturating cotton balls with petroleum jelly.

   - **Friction-Based Methods:**

     - **Bow Drill:** Utilize a bow, spindle, and hearth board to induce friction and produce an ember. Engage in this method regularly to develop expertise.

     - **Hand Drill:** Demands greater expertise and exertion. Generate friction and an ember by rotating a spindle between your palms and a hearth board.

   - **Spark-Based Methods:**

     - **Flint and Steel:** Generate sparks by striking steel against flint. Channel the sparks towards a tinder bundle to initiate a fire. Make sure that the tinder is both dry and finely shredded.

   - **Alternative Methods:**

     - **Magnifying Glass:** Utilize the focused rays of sunlight using a magnifying glass to ignite little, dry materials used for starting a fire. Efficient when the weather is sunny and there is dry tinder available.

- **Battery and Steel Wool:** To short-circuit a battery, touch the wires with steel wool to make sparks. Move the sparks quickly to the tinder.

2. Building and Maintaining a Fire

- **Fire Structure:**
  - **Tinder:** Utilize highly combustible substances such as desiccated foliage, vegetation, or documents. Position a modest quantity in the middle of your fire pit.
  - **Kindling:** Place slender branches or twigs around the tinder to facilitate the construction of the fire. Position them in a relaxed manner to facilitate the circulation of air.
  - **Fuel Wood:** Utilize larger chunks of wood as the major source of fuel. Progressively incorporate larger fragments as the fire expands.

- **Fire Management:**
  - **Airflow:** Optimize airflow by placing the wood in a manner that facilitates the access of oxygen to the fire. Ensure that the fire pit is not excessively crowded.
  - **Maintaining Heat:** Maintain the combustion by adding wood as necessary. Utilize a poker or a stick to manipulate the wood and provide a steady and uniform level of heat.

- **Extinguishing a Fire:**
  - **Dousing:** Utilize water to suppress the fire. Extinguish the fire by pouring water over it and then stirring the ashes to make sure all embers are completely put out.
  - **Smothering:** Smother the fire by applying soil or sand to deprive it of air. Make sure there are no remaining embers that could potentially ignite again.

3. Heat Preservation Techniques

- **Insulation:**
  - **Clothing:** Utilize the strategy of donning numerous layers of clothes to effectively retain and confine body heat. Utilize thermal or insulating materials to augment heat retention.
  - **Blankets:** Utilize blankets or sleeping bags to preserve thermal energy. It is advisable to incorporate additional layers to enhance insulation.

- **Shelter:**
  - **Insulating Materials:** Utilize materials such as pine needles, leaves, or insulated blankets to create a lining on the interior of your shelter. Erect a protective barrier to insulate against the frigid ground.

- **Wind Protection:** Build your shelter to protect against wind and weather. Utilize natural obstacles or supplementary covers to minimize the dissipation of heat.

- **Heat Retention:**
  - **Heat Sources:** Position heat-emitting devices, such as a fire or portable heater, within the shelter. Ensure adequate airflow to prevent the accumulation of carbon monoxide.
  - **Thermal Bottles:** Utilize thermal bottles or insulated containers to maintain the warmth of water or food. Insulate containers by covering them with blankets to enhance heat retention.

- **Ventilation:**
  - **Controlled Airflow:** Ensure sufficient airflow to prevent the formation of condensation and maintain a moisture-free atmosphere. Refrain from producing drafts that may lower the temperature of the shelter.
  - **Ventilation Devices:** Utilize vents or small apertures to facilitate the circulation of fresh air while minimizing heat loss.

4. Emergency Heat Sources

- **Portable Heaters:**
  - **Types:** Utilize battery-powered or propane heaters specifically developed for indoor applications. Adhere to safety protocols in order to avoid carbon monoxide poisoning.
  - **Operation:** Make careful to provide adequate airflow and adhere to the guidelines provided by the manufacturer to ensure safe usage.

- **Heat Packs:**
  - **Chemical Heat Packs:** Utilize disposable heat packs to rapidly generate warmth. Trigger the activation process by vigorously shaking or kneading the object, then securely position it within gloves or pockets.
  - **Reusable Heat Packs:** Utilize heat packs that are designed for several uses and can be heated either by immersing them in hot water or by placing them in a microwave. Ensure that you have many items available for prolonged utilization.

- **DIY Heat Solutions:**
  - **Hot Water Bottles:** Fill containers with heated water and insert them into sleeping bags or garments to provide additional insulation.
  - **Reflective Surfaces:** Utilize reflective blankets or surfaces to redirect heat towards your location. Position them in close proximity to heat sources to amplify warmth.

5. Safety Considerations

- **Fire Safety:**
  - **Fire Watch:** It is essential to consistently keep a fire watch while a fire is actively burning. Always supervise a fire and make sure it is completely put out before departing.
  - **Fire Safety Gear:** Ensure that a fire extinguisher or a container of water or sand is readily accessible in case of emergency.
- **Heat Safety:**
  - **Avoid Overheating:** Monitor the body's temperature to prevent excessive heat buildup. Ensure proper hydration and make necessary adjustments to garment layers.
  - **Ventilation and Carbon Monoxide:** To prevent carbon monoxide poisoning, it is important to have adequate ventilation when utilizing heating devices.

This section offers extensive instructions on methods for igniting fires, constructing and sustaining fires, conserving warmth, and using alternative heat sources in emergency situations. Acquiring proficiency in these abilities is essential for ensuring the retention of heat and the provision of comfort in a survival scenario.

Basic Methods of Self-Defense

During a crisis, it is crucial to possess the knowledge and skills necessary to protect yourself and your loved ones, so assuring safety and effectively handling potential dangers. This section encompasses essential self-defense tactics that are both pragmatic and efficient in diverse situations.

1. Understanding Self-Defense

- **Situational Awareness:**
  - **Assessing Threats:** Cultivate the capacity to identify prospective hazards and evaluate the likelihood of dangers. Take heed of nonverbal cues, conduct, and environment.
  - **Avoidance Strategies:** Emphasize avoidance as the most effective method of self-protection. Minimize exposure to hazardous circumstances and actively avoid engaging in conflicts wherever feasible.
- **Self-Defense Principles:**
  - **Legal Considerations:** Comprehend the legal ramifications of engaging in self-defense. Employ force just in situations where it is essential and commensurate with the level of danger.
  - **Personal Safety:** Concentrate on strategies that empower you to safeguard yourself while minimizing any negative impact on others.

2. **Basic Defensive Stances**

   - **Ready Stance:**

     - **Description:** Assume a stance with your feet positioned at a distance equal to the width of your shoulders, your knees slightly flexed, and your body weight evenly distributed. Maintain a defensive posture by keeping your hands raised in front of you.

     - **Purpose:** The ready posture enables swift responses to incoming assaults while maintaining a stable position for both defensive manoeuvres and retaliatory strikes.

   - **Defensive Posture:**

     - **Description:** Transfer your body's weight to your rear leg while maintaining a defensive stance, with your hands positioned in a protective manner, with palms facing outward. This stance facilitates effortless mobility and defensive manoeuvres.

     - **Purpose:** Offers a solid foundation for protecting oneself from attacks and prepared to retaliate.

3. **Key Defensive Techniques**

   - **Blocking Techniques:**

     - **High Block:** Elevate your forearm to obstruct assaults from above. Maintain a flexed elbow position and utilize the forearm to redirect oncoming attacks.

     - **Low Block:** Lower your forearm to intercept low strikes. Employ your forearm to deflect blows targeted at your lower extremities.

     - **Inside Block:** Employ the underside of your forearm as a defensive measure to obstruct punches or strikes directed towards your midsection. Divert the assault in a direction that is not towards your physical self.

   - **Evading Techniques:**

     - **Sidestep:** Sidestep to evade an imminent assault. This strategy enables you to elude incoming attacks and establish a safe distance from an assailant.

     - **Duck and Weave:** Perform a squatting motion and engage in lateral movement to evade incoming punches or strikes. Employ this strategy to avoid being within the attacker's range of assault.

   - **Counterattacking:**

     - **Straight Punch:** Administer a forceful blow directly to the assailant's abdomen or facial region. Utilize your body mass to generate force and target susceptible regions.

     - **Knee Strike:** Thrust your knee forcefully into the assailant's abdomen or groin area. This strategy is highly efficient for generating distance and rendering the attacker unable to act.

- **Elbow Strike:** Employ your elbow to deliver a forceful blow to the assailant's facial region or upper body. This technique, when used at close range, can be potent and efficient.

4. ## Escape and Evasion

   - **Creating Distance:**
     - **Retreat:** If feasible, disengage from the assailant and relocate to a secure area. Employ impediments or barriers to establish a physical separation and impede the attacker's pursuit.
     - **Escape Routes:** Identify possible exit paths and strategize your movements to swiftly reach a secure location.

   - **Using the Environment:**
     - **Obstacles:** Utilize the objects present in your surroundings, such as furniture or automobiles, to obstruct or impede the progress of an assailant. Strategically position yourself to obtain an advantageous position.
     - **Cover:** Take refuge behind objects or structures to shield yourself from assaults. Utilize cover as a means of protecting yourself and generating chances to flee.

5. ## Personal Safety Tools

   - **Self-Defense Weapons:**
     - **Pepper Spray:** Utilize pepper spray to momentarily disable an assailant. Target the facial area and subsequently execute an escape or employ supplementary defensive manoeuvres.
     - **Personal Alarm:** Carry a portable alarm device to attract attention and discourage prospective assailants. Engage the alarm system to notify and summon assistance from others.

   - **Improvised Weapons:**
     - **Everyday Objects:** Utilize objects such as keys, pens, or small tools as impromptu weapons in a scenario where self-defence is necessary. Target susceptible regions to optimize efficacy.
     - **Defensive Items:** It is advisable to possess goods explicitly created for the purpose of self-defence, such as a tactical pen or a self-defence keychain.

6. ## Self-Défense Training

   - **Practice and Drills:**
     - **Regular Training:** Engage in regular practice of self-defence skills to ensure continued skilfulness. Participate in exercises aimed at enhancing reflexes, coordination, and muscle memory.
     - **Self-Defense Classes:** Enroll in self-defence programs or martial arts instruction to acquire sophisticated methods and practical expertise.

- **Mental Preparedness:**
  - **Confidence:** Enhance your self-defence skills by engaging in rigorous training and consistent practice to cultivate self-assurance. Having a state of mental readiness can enable you to respond efficiently when faced with stressful situations.
  - **Situational Awareness:** Stay alert and aware of your surroundings to anticipate potential threats and respond appropriately.

7. Responding to Physical Confrontations

- **De-escalation:**
  - **Calm Communication:** Employ composed and confident communication techniques to defuse potential disputes. Make an effort to verbally defuse tension before resorting to violent self-defence.
  - **Non-Threatening Body Language:** Adopt a demeanour that does not pose a danger in order to prevent inciting aggression. Ensure that your hands are in plain sight and refrain from making abrupt gestures.
- **Engaging an Attacker:**
  - **Target Vulnerable Areas:** Focus on assaulting susceptible areas such as the eyes, nose, throat, and crotch. These regions are very responsive and can offer possibilities for evasion.
  - **Use Quick, Effective Moves:** Employ techniques expeditiously and resolutely to optimize their efficacy. Minimize extended involvement and favour fleeing whenever feasible.

9. Post-Confrontation Actions

- **Seek Medical Attention:**
  - **Injuries:** Evaluate and provide medical care for any injuries acquired during the altercation. If required, it is advisable to get medical assistance and record any injuries for legal documentation.
- **Report the Incident:**
  - **Authorities:** Notify law enforcement or the appropriate authorities about the occurrence. Please include comprehensive details of the confrontation, including any further steps that were undertaken.
- **Emotional Support:**
  - **Recovery:** If necessary, seek assistance and therapy to receive emotional support. Encountering a confrontation can be distressing and receiving assistance can facilitate the process of healing and managing the situation.

This part provides comprehensive instruction on fundamental self-defence tactics, encompassing defensive postures, crucial defensive and counterattacking manoeuvres, strategies for escaping and evading danger, equipment for personal safety, and recommendations for training. Acquiring proficiency in these approaches will improve your capacity to successfully safeguard yourself in diverse circumstances.

Finding Your Way and Making Help Signals

During a crisis, it is important to have proficient navigation skills and effective signaling techniques in order to locate assistance and guarantee your well-being. This section offers pragmatic advice on navigating strategies, signaling approaches, and utilizing diverse resources to convey your whereabouts and solicit aid.

1. **Navigation Basics**
   - **Understanding Maps and Compasses:**
     - **Maps:** Acquaint oneself with several categories of maps, such as topographic, street, and city maps. Comprehend cartographic symbols, scales, and the skill of interpreting contour lines.
     - **Compass:** Acquire the knowledge and skills necessary to utilize a compass for the purpose of ascertaining direction. Engage in the activity of aligning the compass needle with the magnetic north and adjusting the map's orientation accordingly.
   - **Using GPS Devices:**
     - **GPS Units:** Learn the proper usage of handheld GPS devices or smartphone applications for navigation. Acquire the skills to enter waypoints, adhere to predetermined itineraries, and monitor your precise position.
     - **Battery Management:** Make careful you completely charge your GPS gadgets and, if feasible, bring additional batteries. GPS gadgets can consume a significant amount of battery power and may need the use of alternative power sources.
   - **Landmark Navigation:**
     - **Identifying Landmarks:** Utilize natural or artificial landmarks to establish your position and monitor your advancement. Landmarks encompass various geographical features such as mountains, rivers, buildings, or distinctive attributes.
     - **Creating Landmarks:** Utilize conspicuous objects or different features to establish your location and aid in navigation and the ability to return to certain points.

2. **Signaling for Help**
   - **Visual Signals:**
     - **Signal Mirror:** Utilize a signal mirror to effectively redirect sunlight and draw attention towards oneself. Position the mirror in the direction of the sun and modify the angle to precisely redirect the reflected light towards prospective individuals who can provide assistance.

- **Smoke Signals:** Generate smoke signals by igniting anything such as fresh foliage or damp timber. Utilize sporadic emissions of smoke to indicate a state of emergency.

- **Audible Signals:**
  - **Whistle:** Carry a whistle to emit a resounding and penetrating noise. Produce brief, forceful sounds with the whistle to draw attention and indicate the need for assistance.
  - **Emergency Horn:** Utilize an emergency horn or air horn to generate a resounding and attention-commanding noise. These devices are efficient in areas with high levels of noise or when there is restricted vision.

- **Electronic Signals:**
  - **Emergency Beacons:** Utilize emergency beacons or Personal Locator Beacons (PLBs) to convey distress alerts. These gadgets transmit GPS locations and distress signals to rescue teams.
  - **Cell Phones:** If cellular coverage is accessible, utilize your phone to contact emergency personnel or transmit text messages containing your position. Text communications can be more dependable than voice calls in regions with limited signal strength.

3. **Effective Signaling Techniques**
   - **Signal Patterns:**
     - **SOS:** Indicate distress by employing the universally acknowledged SOS signal pattern, consisting of three short signals, followed by three long signals, and then three short signals again. This pattern is easily identifiable by rescue crews across the globe.
     - **Signal Sequence:** Set up a consistent signaling process using codes or signals that have already been set up. Mix visual, audible, and electronic cues to make it more likely that someone will notice you.
   - **Visibility and Timing:**
     - **Day and Night Signaling:** What you do to signal should depend on the time of day. For daytime signals, use materials that reflect light or are brightly coloured. For nighttime signs, use lights or flares.
     - **Signal Duration:** Keep up your signaling for long amounts of time to improve your chances of being seen. Take breaks to save energy but keep signaling at regular times.

4. **Navigational Tools and Equipment**
   - **Maps and Charts:**
     - **Emergency Maps:** Maintain comprehensive maps of your vicinity and emergency evacuation pathways. Ensure that the maps contain prominent landmarks, reliable water supplies, and any potential hazards.

- **Navigation Charts:** If relevant, utilize specialized navigation charts designed specifically for maritime or aviation purposes. Make sure you have a thorough understanding of the characteristics and representations used in the chart.

- **Compasses and GPS:**

  - **Types of Compasses:** Select compasses of many varieties, including baseplate, lensatic, or digital compasses, according to your specific requirements. Comprehend the operational mechanisms and practical uses of each type.

  - **GPS Accessories:** Bring along GPS accessories such as additional batteries, car chargers, or solar chargers to guarantee the continued functionality of your GPS device.

- **Signaling Devices:**

  - **Flares:** Utilize portable or aerial flares to generate a luminous and conspicuous signal. Flares are highly visible from a considerable distance and are highly useful at capturing attention.

  - **Signal Panels:** Carry conspicuous signal panels or flags that may be readily deployed and observed from a distance. Utilize them to designate your position and augment visibility.

10. Emergency Communication Strategies

- **Establishing Contact:**

  - **Emergency Numbers:** Acquire and retain emergency contact numbers for nearby governmental agencies, rescue services, and assistance organizations. Regularly revise and update contact information.

  - **Communication Plan:** Develop a communication plan with family members or companions. Establish regular check-ins and predetermined signals to ensure everyone can be accounted for.

- **Using Communication Devices:**

  - **Satellite Phones:** When you're in a place that doesn't have cell service, you might want to use a satellite phone to talk. Satellite phones can be used to safely get in touch with emergency services.

  - **Two-Way Radios:** Two-way radios let people in the same group or nearby talk to each other. Make sure that the radios are set to the right channel and that the batteries last long enough.

- **Safety Considerations:**

  - **Avoid Overloading:** Refrain from overwhelming communication channels with unnecessary information. Direct your attention towards effectively communicating essential information and specifying your current whereabouts.

  - **Emergency Frequencies:** Listen to emergency frequencies or channels utilized by rescue teams and emergency agencies. Stay updated on the current progress of search and rescue operations.

NAVY SEALS BUG IN GUIDE

This section offers extensive information on navigation and signaling strategies, encompassing the use of maps, compasses, GPS, and numerous signaling methods. Acquiring proficiency in these abilities will improve your capacity to seek assistance and properly convey your whereabouts during an emergency.

CHAPTER 10
PSYCHOLOGICAL RESILIENCE

Dealing with Stress and Isolation

The effects of isolation and stress on mental health can be substantial during a crisis. This section offers techniques for coping with emotions of isolation, minimizing stress, and preserving emotional well-being.

1. **Understanding Isolation and Stress**

   - **Isolation:**
     - **Definition:** Isolation is the state of being physically separated from people, resulting in emotions of loneliness and separation.
     - **Impact:** Extended alone can lead to psychological anguish, diminished cognitive well-being, and a feeling of powerlessness. Identifying these impacts is the initial stage in dealing with them.
   - **Stress:**
     - **Definition:** Stress is the physiological and psychological reaction that the body experiences in response to perceived dangers or difficulties, which can be expressed through physical, emotional, or mental symptoms.
     - **Impact:** Elevated amounts of stress can hinder cognitive function, disturb sleep patterns, and compromise the immune system. Efficient stress management is crucial for sustaining optimal health.

2. **Managing Feelings of Isolation**

   - **Maintaining Social Connections:**
     - **Virtual Communication:** Utilize technology to maintain communication with friends and family via video calls, social media platforms, or messaging applications. Frequent communication helps reduce feelings of isolation.
     - **Scheduled Check-Ins:** Initiate regular communication with close family and friends. Regular and ongoing communication offers emotional assistance and contributes to the preservation of relationships.

- **Engaging in Community Activities:**
  - **Online Communities:** Engage in online forums, support groups, or virtual events that are relevant to your interests. Interacting with folks who have similar interests and beliefs helps alleviate the sense of being alone.
  - **Volunteering:** Seek virtual or remote volunteer opportunities if they are possible. Assisting others can offer a feeling of purpose and establish a relationship.
- **Self-Care Practices:**
  - **Hobbies and Interests:** Participate in activities that bring you pleasure and a feeling of achievement. Engaging in hobbies can function as a beneficial kind of diversion and enhance one's emotional state.
  - **Mindfulness and Meditation:** Engage in the practice of mindfulness and meditation in order to maintain a state of mental presence and concentration. These strategies aid in the management of feelings of isolation and enhance emotional well-being.

3. Reducing and Managing Stress

- **Identifying Stressors:**
  - **Recognize Triggers:** Identify certain variables that contribute to your stress. Gaining insight into these triggers can assist you in formulating efficient coping mechanisms.
  - **Assess Impact:** Analyse the impact of stressors on your physical, emotional, and mental well-being. Being cognizant of stress reactions allows you to tackle them in a proactive manner.
- **Stress Reduction Techniques:**
  - **Deep Breathing Exercises:** Engage in deep breathing exercises to stimulate the body's relaxation response. Some of the techniques used are diaphragmatic breathing, 4-7-8 breathing, and box breathing.
  - **Physical Exercise:** Integrate consistent physical exercise into your daily schedule. Physical activity stimulates the release of endorphins, which have the ability to enhance mood and alleviate stress.
  - **Progressive Muscle Relaxation:** Utilize progressive muscle relaxation as a technique to alleviate physical tension. This process entails the contraction and subsequent release of various muscle groups.
- **Developing Healthy Routines:**
  - **Daily Schedule:** Establish and adhere to a well-organized daily schedule to ensure consistency and foreseeability. Establishing a regular and predictable routine aid in the effective management of stress and enhances one's overall state of well-being.

- **Sleep Hygiene:** Give priority to implementing effective sleep habits, such as adhering to a consistent sleep routine, establishing a tranquil sleep environment, and refraining from using stimulants prior to going to bed.

4. Building Resilience to Stress

- **Developing Coping Strategies:**

  - **Problem-Solving Skills:** Utilize problem-solving methodologies to effectively tackle and handle stress-inducing factors. Analyse and divide complex problems into smaller, more achievable tasks and concentrate on finding resolutions.

  - **Positive Thinking:** Nurture a constructive mindset by directing your attention towards your strengths, accomplishments, and resolutions instead of fixating on difficulties.

- **Seeking Professional Support:**

  - **Therapy and Counselling:** It is advisable to consider obtaining professional assistance from a therapist or counsellor. Seeking professional counsel can assist you in cultivating effective coping skills and resolving underlying concerns.

  - **Crisis Helplines:** Make use of crisis helplines or mental health resources for instant assistance. These services provide discreet support and guidance during times of stress.

- **Building a Support System:**

  - **Social Network:** Cultivate and sustain a robust support network including of friends, family, and peers. A dependable support system offers both emotional solace and practical assistance.

  - **Community Resources:** Utilize community resources such as support groups, mental health services, and local organizations that provide aid during times of distress.

5. Creating a Stress-Reduction Plan

- **Personal Action Plan:**

  - **Identify Strategies:** Develop a customized strategy that outlines effective ways for reducing stress based on your individual needs. Incorporate strategies such as relaxation exercises, physical exertion, and social interaction.

  - **Set Goals:** Establish attainable objectives for effectively managing stress and enhancing mental well-being. Continuously observe the advancement and make necessary modifications to your strategy.

- **Routine Evaluation:**

  - **Regular Review:** Regularly evaluate and modify your stress management strategy according to your individual requirements and current situation. Remain adaptable and adjust to evolving circumstances.

- **Self-Assessment:** Consistently evaluate your levels of stress and emotional well-being. Utilize self-assessment tools or check-ins to evaluate the efficacy of your strategies.

This section is dedicated to the management of isolation and stress in times of crisis, providing practical techniques for sustaining mental well-being. By doing these behaviours, you can enhance your connectivity, alleviate stress, and cultivate resilience in difficult circumstances.

Developing Mental Fortitude

Psychological resilience is essential for surviving and surmounting the difficulties of a crisis. It requires cultivating resilience, discipline, and a robust attitude to effectively navigate challenging circumstances. This section examines techniques for cultivating and enhancing mental resilience.

1. Understanding Mental Toughness

- **Definition and Importance:**

    - **Mental Toughness:** The capacity to maintain resilience, concentration, and composure in high-stress situations. It requires tenacity, self-assurance, and flexibility when confronted with challenges.

    - **Importance:** Developing mental resilience allows individuals to effectively manage stress, surmount barriers, and sustain optimal performance in the face of adversity. Long-term prosperity and stability are crucial.

- **Core Components of Mental Toughness:**

    - **Self-Confidence:** Having confidence in your capabilities and choices, even when confronted with uncertainty.

    - **Resilience:** Resilience refers to the ability to bounce back from setbacks and persist in the pursuit of goals, even in the face of challenges.

    - **Focus:** The capacity to focus on tasks and objectives, reducing interruptions and sustaining drive.

2. Developing Resilience

- **Adapting to Challenges:**

    - **Embrace Change:** View challenges as opportunities for growth. Adapt to new circumstances by being flexible and open to change.

    - **Learn from Setbacks:** Examine failures and setbacks in order to comprehend the reasons for their occurrence. Utilize these valuable observations to enhance and fortify your strategy.

- **Building Emotional Resilience:**
  - **Positive Self-Talk:** Substitute pessimistic thinking with optimistic affirmations. Motivate yourself and concentrate on your abilities and accomplishments.
  - **Stress Management:** Engage in stress-reduction strategies, such as mindfulness, meditation, and deep breathing, to preserve emotional equilibrium.

3. **Enhancing Self-Discipline**
   - **Setting Goals and Priorities:**
     - **Clear Objectives:** Set clear, quantifiable, and attainable objectives. Well-defined objectives offer guidance and drive.
     - **Prioritization:** Direct your attention towards duties of utmost importance and refrain from delaying or postponing them. Divide things into smaller components and approach them methodically.
   - **Creating Routines and Habits:**
     - **Consistent Routines:** Establish regular rituals that align with your objectives and promote your overall welfare. Consistency fosters the development of discipline and strengthens favourable behaviours.
     - **Healthy Habits:** Integrate practices that foster physical and mental well-being, such as consistent physical activity, nutritious diet, and sufficient rest.

4. **Building Confidence and Self-Efficacy**
   - **Self-Awareness and Strengths:**
     - **Know Your Strengths:** Discover and utilize your personal strengths to enhance self-assurance. Direct your attention towards the areas in which you demonstrate exceptional proficiency and further develop your skills in those specific areas.
     - **Self-Reflection:** Engage in consistent introspection over your accomplishments and advancement. Recognize and commemorate your achievements to strengthen your confidence in yourself.
   - **Facing Fears and Challenges:**
     - **Exposure to Fear:** Incrementally subject yourself to demanding circumstances in order to cultivate self-assurance. Confronting phobias and surmounting challenges improves one's psychological resilience.
     - **Resilient Mindset:** Develop a mentality that perceives problems as chances for personal development. Embrace challenges as an integral component of the path to achieving success.

5. **Maintaining Focus and Motivation**
   - **Concentration Techniques:**
     - **Mindfulness Practice:** Utilize mindfulness strategies to maintain a state of being fully aware and concentrated. Mindfulness aids in diminishing distractions and improving concentration.
     - **Goal Visualization:** Envision the attainment of your objectives in order to maintain motivation and concentration. Envision the sequential actions and resulting consequences to strengthen dedication.
   - **Overcoming Distractions:**
     - **Identify Distractions:** Identify and deal with issues that cause your attention to be redirected. Establish an optimal setting to enhance focus and efficiency.
     - **Manage Time Effectively:** Utilize time management techniques to allocate dedicated intervals for crucial work. Give priority to tasks and refrain from multitasking.

6. **Cultivating a Growth Mindset**
   - **Embrace Learning:**
     - **Continuous Improvement:** Cultivate a mentality that prioritizes the importance of acquiring knowledge and developing oneself. Solicit feedback and regard mistakes as chances for enhancement.
     - **Curiosity and Adaptability:** Maintain a sense of curiosity and receptiveness towards novel ideas and experiences. Embrace flexibility and consistently improve your abilities in response to evolving situations.
   - **Perseverance and Grit:**
     - **Persistence:** Show steadfastness in the pursuit of enduring objectives. Persist in exerting effort and displaying unwavering tenacity in the face of challenges or disappointments.
     - **Resilient Attitude:** Develop a mindset characterized by resilience and perseverance. Embrace difficulties and maintain unwavering dedication to accomplishing your goals.

7. **Building Support Networks**
   - **Social Support:**
     - **Seek Encouragement:** Surround yourself with a circle of folks who provide support, encouragement, and motivation. Establishing positive interactions enhances one's mental resilience.
     - **Build Strong Connections:** Cultivate profound relationships with others who possess similar principles and aspirations. An extensive support system enhances one's emotional resilience.

- **Professional Guidance:**
  - **Seek Mentors:** Interact with mentors or coaches who can provide guidance and assistance. Seeking professional guidance can bolster your resilience and facilitate your personal growth.
  - **Therapeutic Support:** Seek therapy or counselling to receive supplementary assistance in developing mental resilience. Seeking assistance from professionals can offer significant resources and techniques.

This section offers extensive tactics for cultivating mental fortitude, with a particular emphasis on resilience, self-discipline, confidence, concentration, and support networks. By implementing these strategies, you will enhance your capacity to manage emergencies and uphold a resilient and efficient mindset.

Keeping the Spirit Up in Extended Crises

It is essential to uphold morale amid extended crises in order to promote mental fortitude, maintain a positive mindset, and cultivate a feeling of optimism and steadiness. This section provides tactics to maintain high morale, elevate motivation levels, and foster a pleasant atmosphere amid prolonged times of hardship.

8. Understanding the Importance of Morale

- **Definition and Impact:**
  - **Morale:** Psychological and affective condition of a collective or individual, encompassing sentiments of ardour, assurance, and impetus.
  - **Impact on Survival:** Elevated morale fosters improved decision-making, heightened collaboration, and increased resilience. It aids individuals and groups in maintaining concentration and perseverance when confronted with enduring difficulties.

- **Factors Affecting Morale:**
  - **Physical and Emotional Well-Being:** The morale is influenced by sufficient rest, proper nourishment, and emotional support.
  - **Environmental Conditions:** The overall living conditions, encompassing factors such as comfort and safety, have a direct influence on morale.

9. Setting and Maintaining Routines

- **Establishing Daily Routines:**
  - **Consistency:** Establish and adhere to daily schedules to establish organization and anticipate outcomes. Establishing regular and predictable routines is crucial for maintaining a feeling of normality and exerting control over one's life.

- **Routine Activities:** Integrate routine tasks such as eating, physical activity, work, and leisure into your daily timetable. These activities enhance the equilibrium and sustainability of the environment.

- **Incorporating Flexibility:**
  - **Adaptability:** Incorporate adaptability into procedures to account for evolving situations. Modify established procedures as necessary to tackle novel issues or take advantage of new possibilities.

10. Encouraging Positive Communication

- **Fostering Open Dialogue:**
  - **Encourage Expression:** Establish a conducive atmosphere that promotes the ease of sharing personal thoughts and emotions. Effective communication minimizes misinterpretations and fosters confidence.
  - **Active Listening:** Engage in active listening to comprehensively grasp and effectively respond to problems. Recognize and affirm the feelings and experiences of others.

- **Sharing Positivity:**
  - **Positive Reinforcement:** Provide motivation and commendation for exertion and accomplishments. Acknowledge and commemorate little achievements in order to enhance team spirit.
  - **Inspirational Messages:** Share positive and inspiring messages, anecdotes, or quotations to install motivation and inspiration. Positive reinforcement contributes to the maintenance of an optimistic perspective.

11. Maintaining Social Connections

- **Building Community:**
  - **Strengthening Bonds:** Cultivate robust social relationships and bolster support networks. Foster a sense of community by actively participating in group events and promoting joint endeavours.
  - **Group Activities:** Facilitate collective endeavours, such as interactive games, meaningful debates, or collaborative projects, to foster a sense of camaraderie and enhance teamwork.

- **Providing Support:**
  - **Emotional Support:** Provide solace and empathy to anyone in need of emotional assistance. Offer support, compassion, and confidence to individuals who may be facing difficulties.
  - **Help and Assistance:** Be accessible and ready to provide aid or support in tackling tasks or overcoming problems encountered by others. Providing assistance enhances interpersonal connections and reinforces a feeling of unity.

12. Setting Goals and Maintaining Purpose

- **Goal Setting:**

    - **Short-Term Goals:** Establish attainable objectives in the short term to foster a feeling of fulfilment and advancement. Minor achievements contribute to an optimistic perspective.

    - **Long-Term Vision:** Ensure a steadfast dedication to achieving long-term goals and aims. Clarity of vision is essential for maintaining both motivation and direction.

- **Purpose and Meaning:**

    - **Identify Purpose:** Discover and express a clear and profound purpose or significance in the midst of the crisis. Gaining a comprehensive understanding of the wider context or objective can amplify motivation and boost morale.

    - **Focus on Contribution:** Highlight the significance of both individual and collective contributions to the overall endeavour. Acknowledge the manner in which each individual's position contributes to the overall benefit of society.

13. Managing Stress and Maintaining Balance

- **Stress Management Techniques:**

    - **Relaxation Practices:** Utilize relaxation techniques such as deep breathing, meditation, or mindfulness to effectively handle stress and preserve emotional equilibrium.

    - **Physical Activity:** Integrate consistent physical activity into your routine to diminish stress and enhance happiness. Physical activity enhances both physical and mental well-being, as well as the ability to cope with challenges.

- **Work-Life Balance:**

    - **Balance Activities:** Maintain equilibrium between professional responsibilities, relaxation, and recreational pursuits. Refrain from burdening yourself or others with excessive tasks and allocate time for leisure and pleasure.

    - **Self-Care:** Make self-care activities like getting enough sleep, eating well, and taking time for yourself a priority. Self-care is good for both your mental and physical health.

14. Promoting Mental and Emotional Health

- **Encouraging Mental Health Practices:**

    - **Seek Professional Support:** Seek mental health services or professional assistance if necessary. Therapy or counselling might offer supplementary techniques and tactics for sustaining morale.

- - **Engage in Enjoyable Activities:** Engage in pursuits that elicit happiness and promote a state of calmness. Participate in hobbies, leisure, or creative activities to enhance mood and morale.

- **Building Resilience:**

  - **Develop Coping Strategies:** Create and implement effective techniques to manage and overcome stress and challenges. Developing resilience improves your capacity to sustain high spirits under challenging circumstances.

  - **Foster Optimism:** Nurture a constructive and hopeful outlook. Direct your attention towards strengths, possibilities, and potential solutions instead of fixating on issues.

15. Creating a Positive Environment

- **Optimizing Living Space:**

  - **Comfort and Safety:** Ensure that the living environment is comfortable, secure, and promotes well-being. Establish an environment that fosters tranquillity and optimism.

  - **Aesthetics and Personalization:** Customize the surroundings by incorporating significant objects, embellishments, or amenities to boost morale and establish a feeling of homeliness.

- **Encouraging a Supportive Atmosphere:**

  - **Promote Respect and Kindness:** Cultivate an environment characterized by reverence, benevolence, and reciprocal assistance. An optimistic and encouraging environment boosts morale and improves overall well-being.

  - **Resolve Conflicts:** Efficiently address and resolve issues in a timely manner to reduce the detrimental effects of negativity on morale. Ensure ongoing and transparent communication while actively pursuing effective resolutions.

This section offers extensive tactics for sustaining morale throughout extended crises, with a particular emphasis on establishing regularity, effective communication, interpersonal relationships, goal establishment, stress mitigation, and fostering a good atmosphere. Adopting these strategies aids in maintaining motivation, fortitude, and emotional welfare over prolonged difficulties.

CHAPTER 11
MONITORING AND SECURITY SYSTEMS

Setting Up and Keeping an Eye on Home Security Cameras

Home security cameras are an essential element of a comprehensive security system. They offer live monitoring, capture events, and assist in preventing and recording possible dangers. This section provides a comprehensive guide on the necessary procedures for efficiently installing and overseeing home security cameras.

1. **Choosing the Right Cameras**

   - **Types of Cameras:**

     - **Indoor Cameras:** Intended for domestic use. They assist with the surveillance of activities in communal spaces, sleeping quarters, and ports of entrance.

     - **Outdoor Cameras:** Constructed to endure various weather conditions. Perfect for surveillance of outdoor entrances, driveways, and yards.

     - **Wireless vs. Wired:** Wireless cameras provide the advantage of flexible placement and simplified installation, whilst tethered cameras offer steady connectivity and are less prone to interference.

   - **Camera Features:**

     - **Resolution:** Cameras with higher resolutions, such as 1080p or 4K, offer images that are more distinct and contain finer details. Select a resolution that is appropriate for your monitoring requirements.

     - **Night Vision:** Make sure that the cameras are equipped with infrared night vision capabilities in order to capture high-quality footage in low light or complete darkness.

     - **Field of View:** Opt for cameras with a broad field of view to effectively monitor greater regions. Pan-and-tilt cameras provide the ability to modify angles, resulting in improved coverage.

     - **Motion Detection:** Cameras equipped with motion detection have the ability to initiate recording or send warnings when they detect any movement, so assisting in directing attention towards significant occurrences.

2. Planning Camera Placement

- **Determining Key Locations:**
  - **Entry Points:** Place surveillance cameras at primary access points, such as the front and rear entrances. These locations are of utmost importance for surveillance and prevention.
  - **Driveways and Yards:** Install surveillance cameras strategically to monitor driveways, garages, and yards. This facilitates the surveillance of cars and outdoor activities.
  - **Common Areas:** Install surveillance cameras in frequently used areas within the residence to observe and guarantee security.

- **Avoiding Blind Spots:**
  - **Camera Angles:** Arrange the cameras strategically to reduce areas with limited visibility and guarantee complete surveillance. Optimize camera angles to encompass corners and areas that could potentially be overlooked.
  - **Camera Height:** Mount the cameras at an elevation that prevents interference while ensuring unobstructed visibility. Generally, an optimal height range for outdoor cameras is between 8 and 10 feet.

3. Installing Cameras

- **Preparation:**
  - **Tools and Equipment:** Acquire essential equipment such as drills, screws, mounts, and power supply. Adhere to the installation guidelines provided by the manufacturer.
  - **Power Source:** Ensure that the cameras are connected to a reliable power supply. To ensure optimal performance of wireless cameras, it is necessary to have a robust Wi-Fi connection.

- **Installation Steps:**
  - **Mounting:** Mount cameras securely using brackets or mounts. To ensure the endurance of outdoor cameras, it is recommended to utilize waterproof mounts.
  - **Wiring:** For wired cameras, run cables through walls or ceilings to avoid exposure. Use cable management solutions to keep wiring organized.
  - **Alignment:** Optimize camera angles and focus for optimal coverage. Ensure that the camera is positioned in a way that captures the desired areas without any obstructions.

4. **Configuring Camera Settings**

   - **Camera Setup:**

     - **Resolution and Frame Rate:** Optimize camera settings to achieve the best resolution and frame rate according to your specific requirements. Increasing the parameters results in superior quality, but it may necessitate additional storage space.

     - **Motion Detection Zones:** Set up motion detection zones to target certain regions and minimize the occurrence of false notifications. Adjust the sensitivity levels based on the conditions of your surroundings.

     - **Notifications and Alerts:** Activate notifications and alerts for motion detection or system malfunctions. Configure your notification preferences for receiving notifications either through your smartphone or email.

   - **Connecting to a Monitoring System:**

     - **Smartphone Apps:** You can watch cameras from afar by connecting them to apps on your phone. To connect cameras and see live feeds, follow the app's directions.

     - **Computer Access:** Set up a computer or tablet to be able to see camera feeds. Use the software or web tools that came with the camera.

5. **Monitoring and Reviewing Footage**

   - **Real-Time Monitoring:**

     - **Live Feeds:** Live video feeds from cameras can be accessed on computers or through apps for smartphones. Watch what's happening in real time and fix any problems right away.

     - **Camera Views:** You can keep an eye on multiple places at once with multi-camera views. If you need to, you can switch between different camera feeds.

   - **Footage Storage:**

     - **Recording Options:** Select either local storage options such as SD cards or NVR/DVR or opt for cloud storage to save recorded footage. Local storage gives instant accessibility, whereas cloud storage allows remote accessibility and serves as a backup.

     - **Storage Capacity:** Make sure there is enough storage capacity to record and save footage. Frequently monitor storage capacity and oversee recordings to prevent depletion of available space.

   - **Reviewing Recorded Footage:**

     - **Playback:** To rewatch a recorded video, utilize the playback features. Utilize timestamps and motion recognition logs to search for particular occurrences or moments.

- **Incident Documentation:** Record and store video footage of occurrences or behaviours that appear suspicious. Utilize recorded visual material for the purpose of conducting investigations, filing insurance claims, or serving legal needs.

6. Maintaining and Updating Camera Systems

- **Regular Maintenance:**
    - **Cleaning:** Regularly clean camera lenses and housings to maintain good visibility. Clean the camera to eliminate any dust, grime, or obstacles that could impact its effectiveness.
    - **Firmware Updates:** Regularly update camera firmware to maintain optimal security and functionality. Install updates as issued by the manufacturer.

- **Troubleshooting:**
    - **Common Issues:** Tackle prevalent difficulties like as network troubles, image blurriness, or failures in motion detection. If necessary, consult troubleshooting manuals or seek expert help.
    - **Technical Support:** For help with camera issues or configuration concerns, please reach out to technical support. Verify the availability of warranty and support options.

7. Enhancing Camera Security

- **Securing Camera Access:**
    - **Password Protection:** Utilize robust passwords for both camera access and the corresponding accounts. Refrain from using default passwords and periodically change them.
    - **Network Security:** Make sure that the cameras are connected to a secure network that utilizes encryption. Employ firewalls and virtual private networks (VPNs) as a means of safeguarding against unauthorized intrusion.

- **Privacy Considerations:**
    - **Legal Compliance:** Adhere to the legal provisions about monitoring and privacy. Refrain from positioning cameras in areas that are considered private or where individuals anticipate privacy.
    - **Informing Others:** Notify household members and guests on the existence of surveillance cameras. Transparency fosters trust and guarantees that all individuals are cognizant of surveillance.

This guide offers comprehensive instructions for the installation and monitoring of home security cameras. It emphasizes the selection of appropriate cameras, strategic placement, installation, setup, and maintenance. By implementing these techniques, you may improve the effectiveness of your home security system and assure effective surveillance to maintain safety in times of crisis.

Alarms, Motion Sensors, and Integrations with Smart Homes

By using motion detectors, alarms, and smart home technology, you may significantly improve the security of your house. This is achieved through the early detection of invasions, prompt notifications, and the ability to automate various security features. This section provides a comprehensive overview of the fundamental steps involved in establishing and utilizing these systems to enhance the security of your residence.

1. **Understanding Motion Detectors**

   - **Types of Motion Detectors:**

     - **Passive Infrared (PIR):** It detects motion by quantifying alterations in infrared radiation (heat) emitted by objects. Perfect for indoor utilization, particularly in corridors and chambers.

     - **Ultrasonic:** Utilizes acoustic waves to detect movement by quantifying the reflection of sound waves. Applicable for both indoor and outdoor use.

     - **Microwave:** Utilizes microwave transmissions to detect motion by analysing the reflection of those signals. Efficient in expansive regions and capable of penetrating through walls.

     - **Dual Technology:** Integrates passive infrared (PIR) and microwave sensors to minimize false alarms and improve detection precision. Optimal for regions requiring stringent security measures.

   - **Placement and Coverage:**

     - **Strategic Locations:** Place motion sensors strategically in critical locations, such as entrances, driveways, and corridors. Ensure comprehensive coverage of potential entry routes and areas with a high probability of security breaches.

     - **Avoiding Obstructions:** Position detectors to avoid obstructions that might block their field of view, such as furniture or curtains. Ensure clear lines of sight for effective detection.

     - **Adjustable Sensitivity:** Set the sensitivity level of motion detectors according to the environment. Adjust sensitivity to balance between detecting real threats and minimizing false alarms.

2. **Installing and Configuring Alarms**

   - **Types of Alarms:**

     - **Siren Alarms:** Generate high-decibel noises to notify and discourage trespassers. Offered in a range of decibel levels and categories, including indoor, outdoor, and combination alarms.

     - **Silent Alarms:** Transmit notifications to monitoring services or your smartphone silently. Beneficial for inconspicuous notifications and reaching out to law enforcement.

- **Smart Alarms:** Equipped with smart home integration, enabling remote control and monitoring via applications. The features encompass the ability to personalize alerts, set up scheduled activations, and seamlessly integrate with other devices.

- **Installation Tips:**

  - **Placement:** Place alarm sirens strategically in central areas of your home to guarantee optimal audibility. To ensure the security of outdoor alarms, position them at a distance from areas where they may be susceptible to tampering.

  - **Wiring and Power:** To ensure the proper functioning of wired alarms, it is important to ensure that the wiring is correctly installed and connected to the power source. To ensure the proper operation of wireless alarms, it is essential to have batteries installed and in working condition.

  - **Testing:** Regularly conduct tests on alarms to verify their proper functionality. Conduct regular inspections to ensure that alarms are functioning correctly and are effectively integrated.

- **Configuration and Settings:**

  - **Alarm Zones:** Establish distinct alarm zones to designate specific regions within your residence. Adjust the settings for each zone according to its specific security requirements.

  - **Alarm Modes:** Set up alarm modes, such as instant alarms for immediate notifications or delayed alarms for allowing residents to deactivate within a specific timeframe.

  - **Alert Preferences:** Configure alert preferences for various situations, such as receiving messages for activated alarms, updates on system status, and cautions for low battery levels.

3. Integrating with Smart Home Systems

- **Smart Home Hubs:**

  - **Centralized Control:** Utilize a smart home hub to consolidate motion detectors, alarms, and other security devices into a unified system. Hubs facilitate centralized control and automation.

  - **Compatibility:** Verify that the devices are compatible with your smart home hub or system. Verify compatibility with widely used platforms such as Google Home, Amazon Alexa, and Apple HomeKit.

- **Automation and Control:**

  - **Custom Triggers:** Establish automation protocols that activate alarms or alerts in response to motion detector activity or other specified criteria. For instance, the activation of an alarm can occurs when a motion detector detects any movement when the system is in the armed state.

  - **Remote Access:** Utilize smartphone applications to remotely manage and oversee your security system. Gain real-time access to live streams, receive notifications, and make adjustments to settings remotely.

- **Integration with Other Devices:**
  - **Lighting and Locks:** Combine motion detectors and alarms with intelligent lighting and locks. For example, lights can be programmed to activate automatically upon detecting motion, or locks can be activated when an alert is triggered.
  - **Cameras and Sensors:** Integrate security cameras and other sensors to create a comprehensive security solution. For instance, the activation of motion detection can initiate the recording of the camera or the generation of warnings.

4. Monitoring and Response
   - **Real-Time Monitoring:**
     - **Alerts and Notifications:** Get immediate notifications for motion detection, alarm activations, or system failures. Set up notifications to be delivered through smartphone, email, or SMS.
     - **System Logs:** Retrieve system logs to examine the chronological record of past activities, which encompasses activated alarms, instances of motion detection, and interactions performed by users.
   - **Emergency Response:**
     - **Emergency Contacts:** Establish automated alerts to emergency contacts or monitoring services upon activation of an alarm. Ensure timely and immediate response to potential risks.
     - **Response Procedures:** Create response protocols for various situations, such as unauthorized access or system malfunctions. Make sure that all members of the home are knowledgeable about the procedures and know how to appropriately react to alarms.

5. Maintenance and Troubleshooting
   - **Regular Maintenance:**
     - **Battery Replacement:** Replace batteries in motion detectors, alarms, and smart devices as necessary. Frequently monitor the battery levels and promptly replace them before to depletion.
     - **System Updates:** Maintain the latest versions of firmware and software to guarantee the best possible performance and security. Install updates sent by manufacturers.
   - **Troubleshooting Common Issues:**
     - **False Alarms:** Modify the sensitivity levels or relocate the detectors in order to minimize the occurrence of false alerts. Verify that ambient variables, such as the presence of animals or the movement of objects, are not inducing erroneous activations.
     - **Connectivity Problems:** Verify the presence of any connectivity problems with wireless devices. Ensure optimal signal strength and troubleshoot any issues related to interference or communication.

- **System Malfunctions:** Promptly attend to any problems or errors. If necessary, consult troubleshooting manuals or get assistance from technical support.

6. Privacy and Security Considerations

- **Protecting Privacy:**

  - **Data Security:** Implement robust encryption and security measures to safeguard the transmission of data from security cameras, motion detectors, and alarms. Employ robust passwords and establish secure connections to thwart unauthorized entry.

  - **Legal Compliance:** Adhere to the statutes pertaining to monitoring and privacy. Notify household members and visitors on the existence of cameras and detectors.

- **Ethical Use:**

  - **Transparency:** Ensure open and honest communication with household members on the implementation and operation of motion detectors, alarms, and surveillance devices. Ensure comprehensive comprehension of the purpose and advantages of these systems among all individuals.

  - **Respecting Boundaries:** Refrain from installing cameras or sensors in areas that are considered private and where individuals want their privacy to be respected. Adhere to local standards and demonstrate respect for the privacy of your neighbours.

This article offers a thorough explanation on how to install and monitor motion detectors, alarms, and smart home connections. By adhering to these standards, you can improve the security of your residence through efficient detection, prompt notifications, and seamless integration with intelligent home devices.

Establishing a Network for Neighbourhood Watch

A Neighbourhood Watch Network is a community-driven strategy aimed at improving local security by promoting cooperation and attentiveness. By promoting collaboration among residents, exchanging information, and synchronizing actions, you can greatly enhance the security and adaptability of your local community. This section provides a comprehensive guide on how to construct, oversee, and efficiently operate a Neighbourhood Watch Network.

1. Understanding the Basics of Neighbourhood Watch

- **Purpose and Benefits:**

  - **Crime Prevention:** Neighbourhood Watch initiatives seek to prevent crime by enhancing community knowledge and alertness. Active participation in the community can act as a deterrent for future offenders.

- **Community Building:** These activities enhance community bonds and cultivate a feeling of shared assistance and accountability among citizens.
- **Enhanced Communication:** Enhances the exchange of information and communication regarding local security issues and occurrences.

- **Program Structure:**
  - **Coordination:** A Neighbourhood Watch Network often consists of a coordinator or a small team who are responsible for arranging meetings, handling correspondence, and collaborating with local police enforcement.
  - **Participation:** Any neighbours who are interested can engage by attending meetings, reporting any suspicious activity they observe, and providing support for community initiatives.

2. Setting Up the Network

- **Initial Planning:**
  - **Assess Interest:** Assess the level of interest among neighbours by conducting surveys or engaging in informal chats. Identify prospective volunteers who are eager to assume leadership positions.
  - **Define Goals:** Define explicit objectives for the Neighbourhood Watch Network, such as diminishing crime rates, enhancing community involvement, or enhancing emergency readiness.

- **Organizing the Network:**
  - **Designate Leaders:** Designate a coordinator or leadership team to supervise the development and operation of the program. Allocate jobs and duties according to individuals' aptitudes and preferences.
  - **Create a Structure:** Create a network architecture that incorporates neighbourhood zones or blocks to guarantee extensive coverage and effective communication. Designate block captains or area representatives for every zone.

- **Engaging with Law Enforcement:**
  - **Partnerships:** Form alliances with nearby law enforcement authorities. Facilitate the participation of officers in meetings, actively solicit their views, and include them into training and community events.
  - **Resources:** Solicit resources or assistance from law enforcement, such as data on criminal activities, recommendations for ensuring safety, or initiatives aimed at engaging with the community.

3. Promoting the Network

- **Community Outreach:**
    - **Awareness Campaigns:** Utilize leaflets, newsletters, social media platforms, and neighbourhood activities as means to enhance knowledge regarding the Neighbourhood Watch Network and foster active involvement.
    - **Meetings and Events:** Conduct informative gatherings and community functions to present the program, disseminate safety advice, and foster connections among residents.

- **Building Trust:**
    - **Open Communication:** Promote a culture of transparent and courteous communication among residents. Respond to issues and feedback in a positive and inclusive manner.
    - **Inclusivity:** Promote engagement from all members of the community, including various demographics, in order to ensure that the program accurately represents the needs and interests of the entire area.

4. Implementing Effective Strategies

- **Regular Meetings:**
    - **Scheduling:** Conduct regular meetings to address previous occurrences, evaluate safety updates, and strategize future activities. Meetings might occur on a monthly, quarterly, or as-needed basis.
    - **Agenda:** Create a meeting agenda that includes discussions on crime trends, safety tips, and neighbourhood issues. Allocate a specific period for questions and answers as well as for open debate.

- **Reporting Suspicious Activities:**
    - **Guidelines:** Establish explicit instructions for residents on how to promptly and accurately report any dubious behaviours or occurrences. Highlight the significance of reporting information promptly and accurately.
    - **Reporting Channels:** Implement communication channels, such as a designated telephone line, email platform, or online submission form, to enable residents to securely and confidentially provide information.

- **Safety and Crime Prevention Tips:**
    - **Educational Materials:** Disseminate educational resources on safety and crime prevention, including guidelines for securing homes, measures for personal safety, and information on emergency readiness.
    - **Workshops and Training:** Conduct workshops or training sessions focusing on subjects such as first aid, self-defence, and emergency response to improve the skills and knowledge of residents.

5. Utilizing Technology

- **Communication Tools:**

  - **Social media:** Establish exclusive online communities or discussion boards for locals to promptly and effectively exchange updates, notifications, and information.

  - **Mobile Apps:** Utilize neighbourhood safety applications or messaging platforms to promote immediate communication and coordination among members of the network.

- **Surveillance Integration:**

  - **Camera Sharing:** It is advisable to establish a mechanism for distributing surveillance camera recordings to neighbours or law enforcement agencies to assist with investigations and enhance overall security.

  - **Smart Alerts:** Employ intelligent home automation technologies, such as motion sensors or alarms, that can be seamlessly incorporated into the Neighbourhood Watch Network for immediate warnings and surveillance.

6. Maintaining and Growing the Network

- **Continuous Improvement:**

  - **Feedback:** Solicit input from locals regarding the efficacy of the program and potential areas for enhancement. Utilize feedback to make necessary modifications and amplify the program's influence.

  - **Evaluation:** Conduct periodic assessments of the network's functioning, encompassing crime figures, levels of involvement, and general satisfaction. Utilize data to identify achievements and opportunities for improvement.

- **Sustaining Engagement:**

  - **Recognition:** Acknowledge and incentivize individuals who actively participate and volunteer by rewarding them for their valuable contributions. Expressing gratitude contributes to the sustenance of drive and involvement.

  - **Expansion:** Contemplate the possibility of extending the network to other villages or regions that have comparable safety issues. Engage in cooperation with neighbouring communities to exchange resources and tactics.

7. Addressing Challenges

- **Overcoming Resistance:**

  - **Address Concerns:** Address any opposition or apprehension from residents who may be doubtful or hesitant about the initiative. Offer explicit details regarding the advantages and explicitly tackle any apprehensions related to privacy or security.

- **Conflict Resolution:** Resolve any arguments or disagreements inside the network in a diplomatic manner. Foster a constructive and collaborative environment.

- **Handling Incidents:**

    - **Crisis Response:** Create protocols for managing emergencies or incidents within the local community. Ensure that residents are aware of the appropriate actions to take and the appropriate individuals to contact in the event of a crisis.

    - **Coordination:** Ensure seamless collaboration with law enforcement and emergency services during situations. Promote effective and efficient exchange of information and cooperation.

This guide offers a thorough explanation of how to establish and oversee a Neighbourhood Watch Network. It concentrates on the stages of planning, execution, promotion, and upkeep of the network. By implementing a robust security framework that is centred around the community, you may augment the safety and fortitude of your local area.

CHAPTER 12
SHARING OF RESOURCES AND COMMUNITIES

Working Together for Mutual Aid with Neighbours

Engaging in collaborative efforts with neighbours for mutual aid entails establishing a cohesive network wherein community members can provide assistance to one another in times of emergencies or hardships. Efficient collaboration enhances the neighbourhood's ability to withstand and recover from challenges, while also guaranteeing that resources and assistance are distributed fairly. This section examines techniques for promoting collaboration and establishing a reciprocal support system within your community.

1. Building Trust and Relationships

- **Establishing Connections:**

    - **Meet and Greet:** Arrange casual social gatherings to facilitate the introduction of neighbours and initiate the process of establishing interpersonal connections. Events may encompass block parties, potlucks, or community gatherings.

    - **Personal Outreach:** Contact neighbours individually to initiate discussions on mutual help concepts and assess their level of interest. Establishing personal ties can foster trust and promote active engagement.

- **Creating a Supportive Environment:**

    - **Open Dialogue:** Promote a culture of transparent communication among residents to address issues, requirements, and anticipated outcomes. Ensure that all individuals experience a sense of being listened to and appreciated.

    - **Conflict Resolution:** Resolve any problems or disagreements swiftly and in a productive manner. Foster a cooperative and considerate attitude to resolving conflicts.

2. Identifying Community Needs and Resources

- **Assessing Needs:**

  - **Surveys and Discussions:** Administer surveys or organize community gatherings to ascertain the precise requirements and apprehensions of the inhabitants. Collect data on regions where reciprocal assistance would yield the greatest advantages.

  - **Emergency Planning:** Examine potential emergencies or crises that could affect the community and identify the specific forms of assistance that may be necessary.

- **Cataloguing Resources:**

  - **Resource Inventory:** Compile an inventory of resources that community members are willing to offer, including tools, materials, and expertise. Curate a compilation of accessible materials for convenient consultation.

  - **Skill Directory:** Create a comprehensive database of talents and expertise available within the community. Provide details on those who are capable of rendering services such as medical aid, repairs, or counselling.

3. Establishing a Mutual Aid Network

- **Forming a Coordination Team:**

  - **Leadership Roles:** Designate a coordinating team or leaders who will be responsible for overseeing the management of the mutual aid network. Possible roles may consist of a coordinator, communication manager, and resource manager.

  - **Responsibilities:** Specify the duties of each team member, which encompass tasks such as arranging meetings, overseeing resources, and facilitating support.

- **Developing a Plan:**

  - **Mutual Aid Agreement:** Draft a mutual help agreement describing the methods and protocols for distributing resources and providing assistance. Provide instructions for both requesting and offering aid, along with outlining the obligations of all parties involved.

  - **Emergency Response Plan:** Create an emergency response plan that outlines the operational procedures of the mutual help network during times of emergencies. Specify protocols for allocating resources, delivering assistance, and maintaining communication with individuals.

4. **Coordinating Mutual Aid Efforts**
   - **Communication Channels:**
     - **Group Messaging:** Set up group messaging platforms or communication channels for quick and efficient coordination. Options include group chats, email lists, or dedicated apps.
     - **Regular Updates:** Consistently communicate the current progress of mutual assistance endeavours, encompassing any fresh assets, forthcoming activities, or alterations in protocols.
   - **Resource Allocation:**
     - **Request System:** Develop a system to facilitate the process of requesting and distributing resources. Ensure equitable processing of requests and distribution of resources based on necessity.
     - **Logistics:** Oversee the coordination of logistics for the allocation of resources, encompassing the areas of transportation, storage, and monitoring. Ensure the prompt and systematic delivery of resources.

5. **Promoting Active Participation**
   - **Encouraging Involvement:**
     - **Volunteer Opportunities:** Facilitate avenues for citizens to engage in voluntary activities and make valuable contributions to the collective assistance network. Promote engagement in activities such as coordinating events, overseeing resources, or offering support.
     - **Recognition:** Acknowledge and value the contributions made by volunteers and participants. Public recognition and appreciation can serve as a catalyst for ongoing engagement and assistance.
   - **Engaging the Community:**
     - **Outreach Campaigns:** Implement outreach initiatives to bolster the mutual help network and foster active involvement. Utilize fliers, social media platforms, and community gatherings as means to disseminate information.
     - **Feedback Mechanism:** Establish a feedback mechanism to collect comments from community members regarding the efficacy of the mutual help network. Utilize comments to enhance and rectify any concerns.

6. **Ensuring Sustainability and Adaptability**
   - **Maintaining Engagement:**
     - **Ongoing Communication:** Ensure continuous connection with community members to provide them with updates and actively involve them. Consistent updates and periodic check-ins can effectively maintain interest and engagement.
     - **Adaptation:** Ensure readiness to modify the mutual help network in response to shifting requirements, input, and developing situations. Adaptability and promptness are crucial for achieving sustained success.

- **Evaluating Impact:**
  - **Assessment:** Consistently evaluate the influence of the mutual help network on the community. Assess the efficacy of resource allocation, assistance rendered, and overall contentment.
  - **Improvement:** Analyse the network to identify any areas that need improvement and make the required improvements to increase its efficacy and efficiency.

7. **Leveraging External Resources and Support**
   - **Partnerships:**
     - **Local Organizations:** Collaborate with nearby organizations, such as charitable institutions, non-governmental organizations, or religious groups, to enhance the availability of resources and assistance for the collective aid network.
     - **Government and Agencies:** Seek potential partnerships with government agencies or emergency services to gain access to supplementary resources and specialized knowledge.
   - **Grant Opportunities:**
     - **Funding:** Conduct research on grant prospects or funding sources to provide financial support for the mutual aid network. Funding can be utilized to defray expenses associated with resources, equipment, or training.

This guide offers a thorough explanation on how to work together with neighbours to provide assistance to each other. It emphasizes the need of developing trust, recognizing needs and available resources, creating a network, and encouraging active involvement. Through collaboration and resource sharing, communities can strengthen their ability to withstand and support one another during periods of adversity.

Establishing a Community of Support

Establishing a local support network entails constructing a well-organized and cooperative framework within your community to furnish aid, exchange resources, and provide emotional solace in times of emergency. An efficiently structured network can greatly improve community resilience and guarantee that individuals and families obtain the assistance they require. This section provides techniques for creating and sustaining a successful local support network.

1. **Identifying Key Community Members and Leaders**
   - **Recognizing Influencers:**
     - **Local Leaders:** Identify individuals within the local community who possess the ability to exert influence and rally others towards a common cause. This could encompass leaders of local associations, community advocates, or esteemed individuals inside the vicinity.

- **Active Volunteers:** Involve persons who are already engaged in community service or volunteering. Their expertise and dedication can be quite beneficial in establishing a support network.

- **Recruiting Members:**
    - **Diverse Representation:** Ensure that the network incorporates a heterogeneous collection of community people to effectively reflect a wide range of opinions and needs. The presence of diverse elements can augment the efficiency and inclusiveness of the network.
    - **Skill-Based Recruitment:** Enlist individuals possessing specialized skills or experience that are directly applicable to the objectives of the network, such as healthcare experts, educators, or emergency coordinators.

2. Establishing Communication Channels

- **Communication Platforms:**
    - **Online Tools:** Employ online communication technologies, such as group messaging applications (e.g., WhatsApp, Signal), email lists, and social media groups, to ease the exchange of information and coordination.
    - **Offline Methods:** Introduce offline communication strategies for individuals who lack internet connectivity, like community bulletin boards, flyers, and frequent gatherings.

- **Information Sharing:**
    - **Regular Updates:** Please ensure that you regularly provide information regarding network operations, resources, and any changes in procedures. Ensure that all members are notified about the latest updates and forthcoming events.
    - **Feedback Mechanism:** Implement a feedback system to collect input from members of the network and effectively deal with any problems or suggestions for enhancing the network.

3. Organizing Community Meetings and Events

- **Planning Meetings:**
    - **Initial Meeting:** Conduct an inaugural meeting to acquaint individuals with the support network, delineate its goals, and deliberate on assigned tasks and obligations. Take advantage of this opportunity to establish a positive relationship and create a cooperative atmosphere.
    - **Regular Meetings:** Arrange periodic meetings to assess accomplishments, address obstacles, and strategize for future endeavours. Meetings can be conducted either face-to-face or remotely, depending on the preferences and availability of the participants.

- **Hosting Events:**

  - **Awareness Events:** Coordinate and arrange events aimed at increasing knowledge about the support network and the services it provides. Possible events could encompass workshops, informational sessions, or community fairs.

  - **Training Sessions:** Conduct training workshops to instruct community members on pertinent skills, such as first aid, emergency preparedness, or self-defence.

4. Developing a Resource Inventory

- **Resource Collection:**

  - **Donations:** Gather contributions of vital commodities, including provisions, potable water, medical provisions, and equipment. Establish donation stations or organize collection campaigns to acquire resources from members of the community.

  - **Resource Catalog:** Compile a comprehensive inventory of accessible resources, encompassing their precise whereabouts and current state. Ensure that the catalog is periodically updated to accurately represent any changes in the availability of resources.

- **Resource Distribution:**

  - **Distribution Plan:** Create a strategy for allocating resources to those in the community who require assistance. Guarantee that the distribution mechanism is equitable, effective, and easily understandable.

  - **Emergency Kits:** Assemble emergency kits containing vital provisions to be distributed during times of crisis. Kits may consist of items such as medical supplies, non-perishable food, and personal care products.

5. Building a Supportive Environment

- **Fostering Collaboration:**

  - **Teamwork:** Promote synergy and cooperation among members of the network. Highlight the significance of collaboration in order to attain shared objectives and provide mutual assistance.

  - **Mutual Respect:** Foster a culture characterized by reciprocal respect and comprehension. Ensure that all members are respected and their opinions are acknowledged, and immediately resolve any issues or conflicts that arise.

- **Providing Emotional Support:**

  - **Counselling Services:** Provide individuals with access to counselling services or support groups to assist them in managing stress, anxiety, or other emotional difficulties. Please provide details regarding the mental health resources that are currently accessible.

- **Peer Support:** Establish peer support networks to facilitate the exchange of experiences, provision of advice, and mutual encouragement among individuals.

6. Ensuring Continuity and Sustainability

- **Long-Term Planning:**

    - **Strategic Goals:** Create strategic objectives and formulate plans to ensure the long-term viability of the support network. Take into account variables such as financial support, efficient use of resources, and active involvement of the community.

    - **Succession Planning:** Develop a strategy for managing leadership changes and ensuring the ongoing functioning of the support network. Identify those with the potential to become leaders and offer them training and guidance.

- **Evaluation and Improvement:**

    - **Performance Review:** Conduct periodic evaluations of the support network's performance to gauge its efficiency and pinpoint areas that can be enhanced. Collect input from those who are part of the organization and those who have a vested interest in its activities.

    - **Adaptation:** Be ready to modify the network's plans and operations in response to shifting requirements, feedback, and developing conditions. Consistently strive to improve the network's influence and effectiveness.

7. Collaborating with External Partners

- **Local Organizations:**

    - **Partnerships:** Establish collaborative alliances with nearby organizations, such as non-governmental organizations, philanthropic institutions, and religious associations, in order to enhance the network's resources and assistance.

    - **Shared Resources:** Utilize external resources, such as financial support, educational programs, or specialized knowledge, to improve the network's capacities and efficiency.

- **Government and Agencies:**

    - **Coordination:** Collaborate with local government agencies and emergency services to obtain supplementary resources and assistance. Cultivate connections with essential contacts to facilitate cooperation during emergencies.

    - **Grants and Funding:** Investigate potential sources of grants or funds to provide financial support for the network's operations and goals. Submit applications for appropriate awards in order to obtain financial assistance.

This guide offers a thorough overview of how to form a local support network, with a particular emphasis on cultivating relationships, setting up effective communication channels, coordinating meetings and activities, compiling a resource inventory, promoting collaboration, and guaranteeing long-term viability. Communities can bolster their ability to withstand and successfully address emergencies by establishing a robust and encouraging network.

Trading and Safe Resource Sharing

Engaging in bartering and resource sharing might prove to be efficacious methods of acquiring essential goods and services in times of crisis, particularly when conventional systems may encounter disruptions. Nevertheless, it is imperative to approach these operations with prudence in order to guarantee safety, equity, and efficacy. This section offers tactics for engaging in the exchange of goods and services and distributing resources securely within your nearby community of assistance.

1. **Understanding the Principles of Bartering**

   - **Bartering Basics:**

     - **Definition:** Bartering is the act of directly exchanging commodities or services for other goods or services, without the use of money. It can be a beneficial strategy for obtaining necessary things when cash is unavailable.

     - **Value Assessment:** Evaluate the worth of the things or services you provide and those you are looking for. Ensure that both sides regard the exchange as just and balanced.

   - **Negotiation Skills:**

     - **Clear Communication:** Ensure clear and concise communication regarding the specifics of your offerings and requirements. Provide accurate and transparent information regarding the excellence and amount of goods or services.

     - **Flexibility:** Be ready to engage in negotiations and make modifications to the terms of the transaction if needed. Adopting a flexible approach can facilitate the achievement of mutually advantageous agreements.

2. **Establishing Bartering Guidelines**

   - **Setting Standards:**

     - **Value Guidelines:** Develop guidelines for assessing the value of items and services. Consider factors such as condition, scarcity, and demand.

     - **Fair Exchange:** Make sure that all trading deals are honest and fair. Don't take advantage of other people or accept deals that aren't good for both of you.

- **Documentation:**
  - **Record-Keeping:** When you trade something for something else, write down the terms of the deal and how to reach the other person. Keeping records can help settle disagreements and keep track of deals.
  - **Receipts:** Please furnish or ask for receipts for substantial transactions. Receipts can function as evidence of the transaction and aid in upholding responsibility.

3. Sharing Resources within the Community

- **Resource Sharing Principles:**
  - **Equity:** Ensure that resources are distributed fairly and evenly among all members of the community. Give priority to individuals with the most urgent needs and refrain from hoarding or distributing resources unequally.
  - **Transparency:** Ensure clear and open communication regarding the accessibility of resources and the standards for their distribution. Articulate the precise method by which resources will be distributed.

- **Resource Pools:**
  - **Creating Pools:** Create resource pools where community members can contribute or loan stuff. Pools may encompass provisions such as sustenance, medical resources, implements, or apparatus.
  - **Management:** Allocate persons or teams to oversee the resource pools, guaranteeing efficient tracking, maintenance, and distribution of things.

4. Ensuring Safety in Bartering and Sharing

- **Health and Hygiene:**
  - **Sanitation:** Make sure that all objects being exchanged or shared are both hygienic and secure. Adhere to appropriate cleanliness protocols to mitigate the transmission of illnesses.
  - **Safety Checks:** Prior to exchanging or sharing, thoroughly examine objects for any potential safety concerns. For instance, ensure to verify the expiration dates on food items or thoroughly examine equipment for any signs of damage.

- **Security Measures:**
  - **Verification:** Authenticate the identities of those participating in the exchange of goods or sharing of resources. Refrain from engaging with unfamiliar or dubious individuals.
  - **Safe Exchanges:** Whenever feasible, carry out transactions in secure and impartial venues. To facilitate home-based exchanges, it is advisable to schedule meetings with reliable individuals and communicate your intentions to others.

5. Establishing Cooperation and Trust in the Community

- **Fostering Trust:**
  - **Reliability:** Exhibit dependability and consistently fulfil commitments. Trust is a crucial component for achieving successful exchanges and sharing among members of the community.
  - **Respect:** Ensure that all participants are treated with dignity and impartiality. Cultivate favourable relationships to foster ongoing collaboration and assistance.

- **Community Engagement:**
  - **Involvement:** Facilitate dialogues among community members regarding the practice of bartering and the sharing of resources. Engage them in the process of decision-making and planning to increase their engagement and garner support.
  - **Feedback:** Collect input from community members regarding their encounters with bartering and resource sharing. Utilize feedback to enhance processes and effectively resolve any difficulties.

6. Legal and Ethical Considerations

- **Legal Compliance:**
  - **Regulations:** Familiarize yourself with the local norms and laws pertaining to bartering and resource sharing. Ensure that all activities adhere to legal regulations.
  - **Permits:** If necessary, acquire all requisite permits or clearances for hosting large-scale resource sharing or bartering events.

- **Ethical Practices:**
  - **Ethics:** Adhere to ethical principles when engaging in bartering and resource-sharing activities. Refrain from participating in activities that may be perceived as exploitative or unjust.
  - **Transparency:** Ensure transparency in transactions and ensure that all parties are fully informed of the terms and conditions of the exchange.

7. Managing Disputes and Conflicts

- **Dispute Resolution:**
  - **Resolution Mechanism:** Create a system to address and resolve any issues or conflicts that may occur throughout the process of trading or sharing resources. Contemplate the utilization of mediation or arbitration as potential resolutions.
  - **Communication:** Resolve any conflicts that arise in a timely and transparent manner. Engage in effective communication with all parties concerned in order to comprehend their viewpoints and reach a just settlement.

- **Preventive Measures:**
    - **Clear Agreements:** Establish unambiguous agreements for all transactions to minimize misinterpretations. Provide specific information, such as detailed descriptions of the items, numbers involved, and the terms under which the exchange will take place.
    - **Guidelines:** Create a set of protocols for managing disputes and disagreements. Distribute these rules to community members to guarantee a uniform approach.

This guide offers a thorough overview of the safe practices involved in bartering and resource sharing. It emphasizes the importance of principles, guidelines, safety precautions, trust-building, legal and ethical factors, and conflict resolution. By implementing these measures, you can promote efficient and fair distribution of resources within your community.

CHAPTER 13
SUPERIOR HOUSE DEFENSE

Strengthening the Structural Integrity of Your House

Strengthening the structural integrity of your home is essential for improving its ability to withstand physical attacks, natural disasters, and other potential dangers. Reinforcing the framework of your residence can guarantee its safety and stability during severe circumstances. This section offers tactics and procedures for strengthening several elements of your home's framework.

1. **Foundation and Walls**

   - **Foundation Reinforcement:**

     - **Inspection:** Commence by doing a comprehensive examination of your dwelling's foundation to detect any preexisting vulnerabilities or harm. Inspect for fissures, subsidence, or any other indications of structural problems.

     - **Piers and Supports:** Implement supplementary piers or supports to fortify the foundation and avert any movement or sinking. It is advisable to utilize steel or concrete piers for increased durability.

     - **Foundation Repair:** Rectify any fissures or impairment in the foundation by employing suitable restoration techniques, such as epoxy injections or concrete resurfacing.

   - **Wall Reinforcement:**

     - **Steel Studs:** Substitute or augment wooden studs with steel studs in crucial sections of your residence. Steel studs offer enhanced durability and resilience against both impact and pressure.

     - **Reinforced Concrete:** It is advisable to use reinforced concrete or masonry for both the external and internal walls. Reinforced concrete can increase the wall's resistance to external forces.

     - **Bracing:** Implement diagonal bracing or shear panels to fortify walls and mitigate the risk of collapse or deformation in times of emergency.

2. Roof and Ceiling Strengthening

- **Roof Reinforcement:**

    - **Trusses and Rafters:** Enhance or strengthen roof trusses and rafters by adding further support or metal braces. Verify the roof structure's ability to endure substantial loads, such as snow or debris.

    - **Roof Decking:** Utilize roofing materials that are resistant to impact or reinforced to provide protection against extreme weather conditions. Take into account the option of using metal roofing or high-strength shingles.

    - **Hurricane Straps:** Attach hurricane straps or ties to fasten the roof to the walls, so safeguarding it against uplift or damage caused by powerful winds.

- **Ceiling Reinforcement:**

    - **Cross-Bracing:** Enhance the stability and prevent drooping or collapse of the ceiling joists by including cross-bracing. Utilize metal or timber bracing elements as suitable.

    - **Support Columns:** Install supplementary support columns or beams in regions experiencing substantial loads or frequent traffic. Make sure that these supports are securely fastened and positioned correctly.

3. Doors and Windows

- **Reinforcing Doors:**

    - **Impact-Resistant Doors:** Install doors that are designed to withstand impact or strengthen existing doors by adding reinforcement. It is advisable to utilize metal or composite doors that have strengthened frames.

    - **Secured Hinges:** Utilize robust hinges and fasten them securely with tamper-resistant screws. Strengthen the door frame to deter unauthorized access.

- **Reinforcing Windows:**

    - **Window Film:** Enhance the durability and resilience of the glass by applying security window film, which will improve its ability to withstand impact and prevent shattering. Using film can effectively reinforce cracked glass, making it more difficult for unauthorized access.

    - **Reinforced Frames:** Replace the window frames with metal or strengthened materials. Make sure that frames are firmly fastened and designed to withstand attempts at unauthorized access.

    - **Shutters and Bars:** Enhance the security of your windows by installing security shutters or metal bars. Select shutters that are readily deployable and removable as required.

4. Structural Enhancements for Vulnerable Areas

- **Basements and Cellars:**

    - **Reinforced Walls:** Strengthen the basement walls with steel or concrete to avoid potential collapse or floods. Implement sump pumps and waterproofing techniques to safeguard against water damage.

    - **Secure Access:** Strengthen the entrances to the basement by installing sturdy doors and reliable locks. Make sure to secure basement windows with either bars or reinforced glass.

- **Attics:**

    - **Strengthening Roof Supports:** Strengthen the structural integrity of the attic roof supports to avoid any potential sagging or complete collapse. Install further bracing or support beams as necessary.

    - **Ventilation:** To mitigate the danger of structural damage caused by temperature variations, it is essential to maintain enough ventilation in the attic to prevent the buildup of heat.

5. Enhancing Overall Structural Integrity

- **Load-Bearing Capacity:**

    - **Assessing Loads:** Assess the ability of important structural components, such as beams, columns, and joists, to support weight. Implement essential enhancements to accommodate increased weight or strain.

    - **Reinforcement Techniques:** Utilize reinforcing methods such as steel plates, brackets, or additional framework to augment the load-bearing capability of structural components.

- **Seismic Reinforcement:**

    - **Earthquake Bracing:** Implement seismic-resistant bracing and anchoring methods to safeguard against earthquakes. Utilize foundation bolts and shear walls for the purpose of stabilizing the construction.

    - **Retrofitting:** It is advisable to enhance the earthquake resilience of your home by upgrading it with seismic modifications. Seek advice from a structural engineer for recommendations.

- **Fire Resistance:**

    - **Fire-Resistant Materials:** Utilize fire-resistant materials for essential structural components, such as fire-rated doors, walls, and ceilings. Administer flame-retardant coatings to susceptible regions.

    - **Fire Barriers:** Implement fire barriers or fire-resistant insulation to prevent the propagation of fire. Ensure that fire barriers are adequately sealed and regularly maintained.

6. Professional Assessment and Installation

- **Consulting Experts:**

    - **Structural Engineer:** Seek the expertise of a structural engineer to evaluate the current structural soundness of your home and obtain suggestions for strengthening it. Experts can offer specialized knowledge and advice on modern fortification techniques.

    - **Contractors:** Engage proficient contractors to implement reinforcing measures. Verify that contractors possess the necessary licenses and qualifications for the specific tasks they are undertaking.

- **Quality Assurance:**

    - **Inspections:** Perform routine examinations of reinforced regions to verify their proper functionality. Resolve any issues or repair any damage swiftly.

    - **Maintenance:** Conduct regular maintenance on reinforced elements to ensure they remain in excellent condition. Adhere to the manufacturer's guidelines and follow professional advice for maintenance.

This guide offers a thorough explanation of how to strengthen the structural integrity of your home. It includes instructions on reinforcing the foundation and walls, strengthening the roof and ceiling, enhancing doors and windows, and making structural changes to vulnerable places. By applying these tactics, you can greatly enhance your home's ability to withstand and defend against a wide range of potential dangers.

Bulletproofing and Blast Protection Options

Strengthening the security of your home against ballistic threats and explosive blasts requires the implementation of sophisticated security systems specifically built to endure catastrophic impacts. This section examines various strategies for reinforcing your home to withstand high-level threats, such as bullet-proofing and blast protection.

1. Bulletproofing Your Home

- **Bulletproof Glass:**

    - **Types of Bulletproof Glass:** Utilize ballistic-resistant glass specifically engineered to endure various degrees of firearm impacts. There are various types of materials used, such as polycarbonate, laminated glass, and glass-clad polycarbonate.

    - **Installation:** Implement the installation of shatterproof glass in both windows and doors. Make sure that the glass is correctly installed and tightly sealed to optimize protection. If complete window replacement is not possible, it is advisable to enhance the existing windows by installing ballistic-resistant film.

- **Bulletproof Doors:**
  - **Materials:** Select doors constructed from materials that possess ballistic resistance, such as reinforced steel or composite materials. Verify that doors are subjected to rigorous testing to fulfil precise ballistic criteria.
  - **Reinforcement:** Strengthen doors by adding supplementary ballistic panels or inserts. Ensure that frames are sturdy and able to bear impact.
- **Ballistic Panels:**
  - **Application:** Mount ballistic panels on walls and doors in strategic locations within the residence. Panel options encompass a range of dimensions and depths, and they can be affixed to preexisting surfaces.
  - **Integration:** Incorporate ballistic panels into the current security infrastructure, including locks and alarms. Ensure that panels are installed firmly and without any spaces.
- **Safe Rooms:**
  - **Bulletproof Construction:** Create secure chambers employing ballistic materials, such as fortified walls, doors, and windows. Verify that the safe room has the ability to withstand high-velocity projectiles.
  - **Shelving and Equipment:** Install bullet-resistant shelving and storage units in the safe room to store necessary goods. Ensure the inclusion of communication equipment and first aid kits.

2. Blast Protection for Your Home

- **Blast-Resistant Windows:**
  - **Types of Windows:** Utilize blast-resistant windows constructed from laminated glass or polycarbonate materials for installation. These windows are specifically engineered to absorb and disperse the energy resulting from explosions.
  - **Installation:** Ensure the proper installation of blast-resistant windows by using sturdy frames and anchoring mechanisms. Please contemplate enhancing the structural integrity of window apertures by adding supplementary support.
- **Blast-Resistant Doors:**
  - **Materials and Design:** Utilize blast-resistant doors constructed from reinforced steel or composite materials. These doors are specifically engineered to endure and resist high-pressure explosions, thereby obstructing any attempts to gain access.
  - **Sealing:** Make sure that doors are adequately sealed to avoid the infiltration of pressure waves into the structure. Place blast-resistant gaskets and seals along the edges of the doors.

- **Reinforced Walls and Ceilings:**
  - **Reinforcement Techniques:** Strengthen walls and ceilings with blast-resistant materials, such as reinforced concrete or steel. Enhance blast resistance by applying more layers or coatings.
  - **Structural Support:** Utilize structural reinforcements, such as steel beams or braces, to enhance the strength of walls and ceilings. Make sure that the supports are firmly fastened to resist the impact of explosions.
- **Blast Barriers:**
  - **Placement:** Install blast barriers around key areas of the home, such as entryways and windows. Barriers can include reinforced walls, fences, or other structures designed to absorb blast energy.
  - **Construction:** Build barriers utilizing materials such as reinforced concrete or steel. Ensure that barriers have sufficient height and width to offer good protection.

3. **Integration of Bulletproof and Blast Protection**
   - **System Coordination:**
     - **Layered Protection:** Integrate bullet-proofing and blast protection measures to establish a thorough security system. Ensure the smooth integration of all protection elements to achieve optimal security.
     - **Design Considerations:** Take into account the design and arrangement of your property in order to properly incorporate bulletproof and blast protection. Ensure the absence of any potential vulnerabilities that could be exploited by threats.
   - **Testing and Maintenance:**
     - **Regular Testing:** Regularly test the bulletproof and blast protective features to verify their efficiency. Administer drills or simulations to evaluate the effectiveness of protective measures.
     - **Maintenance:** Conduct regular maintenance on bulletproof and blast protection components. Inspect for any signs of damage, wear, or deterioration and perform any required repairs or replacements.

4. **Professional Consultation and Installation**
   - **Consulting Experts:**
     - **Security Specialists:** Seek advice from security experts or contractors who possess expertise in the field of bullet-proofing and blast mitigation. They possess the knowledge and can offer suggestions for successful execution.
     - **Custom Solutions:** Collaborate with professionals to create and implement personalized bulletproof and blast protection systems that are specifically designed to address the unique requirements and susceptibilities of your residence.

- **Certification and Standards:**
  - **Compliance:** Verify that all bulletproof and blast protection materials and systems adhere to applicable requirements and certifications. Seek out products that have undergone rigorous testing and obtained certification from reputable organizations.

This article offers a thorough overview of the various alternatives available for reinforcing your home's security against ballistic and explosive hazards, with a focus on bullet-proofing and blast protection measures. By incorporating these sophisticated strategies, you can greatly enhance your home's ability to withstand and safeguard against hazardous circumstances.

Establishing Covered Safe Areas

Constructing covert sanctuaries within your residence offers an extra level of safeguarding, enabling you to conceal and defend yourself in case of a crisis. These concealed regions are essential for guaranteeing security during incursions or other pivotal circumstances. This section provides guidelines for creating and executing successful hidden secure areas.

1. Identifying Potential Locations

- **High Traffic Areas:** It is advisable to install hidden compartments in parts of the house that are unlikely to be inspected by trespassers, such as behind bookcases, inside bulky furniture items, or beneath staircases.
- **Unused Spaces:** Make use of underutilized or less conspicuous locations, such as attics, basements, crawl spaces, or spaces hidden behind false walls, to provide disguised secure zones.
- **Accessible Yet Hidden:** Select areas that are conveniently reachable for you and your family, while being challenging for others to uncover. Make sure that these areas are discreetly concealed yet readily accessible to you in an emergency.

2. Designing Concealed Safe Spaces

- **Hidden Entrances:**
  - **Bookshelves and Panels:** Implement concealed entrances or compartments positioned discreetly behind bookcases, wall panels, or cabinetry. Make sure that the method of accessing the hidden area is inconspicuous and not readily noticeable.
  - **False Walls:** Create deceptive partitions or portable panels that may be easily opened to expose the hidden secure area. Ensure that the deceptive partition easily integrates with the surrounding interior design.
  - **Secret Compartments:** Construct partitions within preexisting furnishings, such as beds, sofas, or workstations. These compartments can serve as concealment spaces for important things or as hiding places.

- **Concealment Techniques:**
  - **Camouflage:** Employ camouflage techniques to seamlessly integrate the entry of the hidden room with its environment. To enhance the camouflage of hidden doors or panels, one can employ paint, wallpaper, or upholstery.
  - **Disguised Mechanisms:** Utilize covert devices, such as secret latches, electronic locks, or magnetic catches, to access hidden compartments. Ensure that these systems are both dependable and fortified.

3. **Equipping Concealed Safe Spaces**
   - **Safety Essentials:**
     - **Emergency Supplies:** Outfit the designated secure area with vital emergency provisions, such as potable water, non-perishable sustenance, a medical kit, and fundamental equipment. Make sure to store goods in sturdy, waterproof containers.
     - **Communication Devices:** Ensure that the safe space contains a battery-operated or hand-crank emergency radio as well as a properly charged phone or communication device. It is advisable to incorporate a two-way radio for effective communication with individuals outside the secure area.
   - **Comfort and Survival Items:**
     - **Blankets and Sleeping Gear:** Enhance the livability of the safe environment by incorporating additional comfort goods such as blankets, sleeping bags, or other such amenities. If possible, it would be advisable to include a compact and easily transportable warming or cooling appliance.
     - **Sanitation Supplies:** Keep essential sanitary supplies in stock, like disinfecting wipes, toilet paper, and a compact portable toilet or waste disposal bags.
   - **Lighting and Visibility:**
     - **Emergency Lighting:** Place a compact, battery-powered or manually operated flashlight and extra batteries within the hidden compartment of the safe. It is advisable to utilize glow sticks or emergency candles as supplementary sources of illumination.
     - **Reflective Materials:** Utilize reflective materials or emergency signaling devices to augment visibility in order to draw the attention of rescuers.

4. **Access and Security**
   - **Access Points:**
     - **Multiple Entrances:** It is advisable to establish many hidden entry methods to the secure area, enabling more adaptability during emergency situations. Verify that all entry points are both secure and operational.

- **Training:** Provide comprehensive training to all family members on the proper procedures for swiftly and securely accessing the hidden safe area. Carry out practice drills to ensure that all individuals are acquainted with the procedure.

- **Security Measures:**

  - **Locking Mechanisms:** Employ robust locking systems, such as deadbolts or electronic locks, to safeguard the concealed protected area from illegal entry. Ensure that the locks are dependable and resistant to tampering.

  - **Emergency Exit:** Develop a contingency plan for a discreet safe location to ensure a swift and secure exit in the event of an emergency or the need to evacuate. Make sure that the exit is unobstructed and easy to reach.

5. Maintenance and Inspection

- **Regular Checks:**

  - **Inspection:** Perform routine inspections of the hidden secure area to verify its structural integrity and confirm the appropriate operation of the entry mechanisms. Inspect for any indications of wear or harm.

  - **Replenishment:** Regularly restock emergency provisions and replace any products that have expired or been destroyed. Ensure that the designated area is consistently supplied with necessary materials.

- **Upgrades:**

  - **Improvement:** Contemplate enhancing or altering the hidden secure area in response to evolving requirements or emerging dangers. Keep yourself updated on the latest developments in safety and security to improve the protection of your concealed area.

6. Professional Assistance

- **Consulting Experts:**

  - **Security Professionals:** Seek guidance from security experts or specialized contractors in concealed safe places to develop and implement top-notch, efficient solutions. They may offer expert advice on the optimal materials and methodologies suitable for your specific requirements.

  - **Custom Solutions:** Collaborate with specialists to design personalized hidden secure areas that are customized to fit the style of your residence and meet your individual security needs.

This book offers a thorough explanation on how to create hidden secure areas in your home, including information on design, equipment, access, security, and upkeep. By applying these tactics, you may improve the safety of your home and guarantee a secure sanctuary during times of need.

CHAPTER 14
ADJUSTING TO SHIFTING CIRCUMSTANCES

Adapting Your Strategy to Various Situations

It is crucial to modify your survival plan to different situations in order to efficiently handle a broad spectrum of probable emergencies. This section provides ways for adapting your strategy in response to various types of crises, ensuring that you can effectively address each scenario and uphold safety and security.

1. **Natural Disasters**

   - **Earthquakes:**

     - **Immediate Actions:** Ensure the stability of bulky furniture and equipment by fastening them securely to prevent them from toppling over. Take proactive steps to strengthen safety precautions and ensure that emergency supplies are easily reachable in anticipation of potential aftershocks.

     - **Post-Quake Adjustments:** Inspect for any structural damage and provide repairs as needed. Evaluate the stability of your residence and address any required repairs to guarantee safety.

   - **Floods:**

     - **Preparation:** Raise essential supplies and equipment to avoid water damage. Identify and have ready substitute water sources as a precautionary measure in the event of contamination.

     - **Response:** Seek shelter at a higher elevation within your residence or evacuate if deemed essential. Employ water filtration and purification techniques to guarantee the consumption of potable water.

   - **Hurricanes and Tornadoes:**

     - **Preparation:** Strengthen windows and doors to endure strong winds. Construct an inner sanctuary or fortified shelter within the confines of your residence.

- **Response:** Take refuge in the specific area designated as a safe space. Following the storm, evaluate the extent of the damage and promptly attend to any urgent requirements, such as power disruptions and water supply problems.

2. **Man-Made Disasters**
   - **Chemical or Biological Incidents:**
     - **Preparation:** Acquire essential protective equipment like as masks, gloves, and decontamination supplies for stock. Develop a strategy to effectively secure your residence in order to avoid any form of pollution.
     - **Response:** Adhere to the prescribed emergency procedures for decontamination and take precautions to prevent any contact with harmful substances. Utilize air and water filtering systems for the purpose of eliminating impurities.
   - **Power Outages:**
     - **Preparation:** Ensure the presence of alternative power sources, such as generators or solar panels, to provide backup power. Acquire a sufficient supply of non-perishable food and water to maintain yourself in the event of power or utility failures.
     - **Response:** Utilize alternate energy sources to sustain essential systems. Optimize energy consumption by giving priority to indispensable appliances and lighting.
   - **Civil Unrest or Terrorism:**
     - **Preparation:** Strengthen residential security protocols and establish a comprehensive strategy for implementing lockdown or evacuation procedures. Ensure you are aware of possible dangers and consistently stay in contact with local law enforcement.
     - **Response:** Adhere to security standards and remain indoors if it is required. Utilize the security measures in your residence to safeguard against unauthorized individuals and maintain a state of constant alertness.

3. **Medical Emergencies**
   - **Pandemics:**
     - **Preparation:** Ensure that you have an ample supply of medical essentials, like masks, hand sanitizer, and antiviral drugs. Enforce sanitary measures and establish a contingency plan for isolation if necessary.
     - **Response:** Adhere to the instructions provided by public health authorities and quarantine yourself if you have contracted an infection.

       Observe symptoms and obtain medical attention if needed.

- **Injury or Illness:**
  - **Preparation:** Assemble a comprehensive first aid kit and familiarize yourself with basic medical procedures. Ensure that you have access to essential medications.
  - **Response:** Administer initial medical assistance and promptly seek professional medical care if necessary. Revise your plan according to the seriousness of the injury or illness and the resources that are now accessible.

4. **Social or Economic Disruptions**
   - **Economic Downturns:**
     - **Preparation:** Develop a comprehensive financial strategy and allocate funds specifically for unforeseen circumstances. Accumulate necessary resources to lessen the effects of economic volatility.
     - **Response:** Review your budget and make necessary adjustments to your spending. Optimize the utilization of your resources and investigate alternate avenues for generating cash or obtaining support.
   - **Supply Chain Disruptions:**
     - **Preparation:** Accumulate necessary items and explore other sources for provisions. Create a comprehensive strategy for effectively managing resources and, if needed, engaging in bartering.
     - **Response:** Allocate and distribute your resources according to their availability. Utilize local resources or community support to tackle shortages.

5. **Personal Emergencies**
   - **Family Health Issues:**
     - **Preparation:** Ensure that you have an up-to-date record of your medical history and contact details for your healthcare professionals. Ensure that you have an adequate supply of essential medications and medical supplies.
     - **Response:** Promptly address health issues and adapt your plan according to specific medical requirements. Consult with a healthcare professional for necessary medical advice and treatment.
   - **Loss of Shelter or Home:**
     - **Preparation:** Ensure you have a contingency plan in place for other possibilities for refuge. Ensure that emergency supplies are stored in a portable and easily accessible manner.
     - **Response:** Make use of interim shelter alternatives and evaluate long-term housing remedies. Modify your plan to meet alterations in living arrangements.

6. **Preparing for Unknown Scenarios**
   - **Flexible Planning:**
     - **Scenario Planning:** Create versatile strategies that can be adjusted to various types of emergencies. Examine different situations and possible reactions.
     - **Regular Updates:** Consistently evaluate and revise your strategies to accommodate fresh data or evolving conditions. Remain knowledgeable about developing dangers and adapt accordingly.
   - **Training and Drills:**
     - **Scenario Drills:** Engage in frequent exercises for various situations to rehearse and enhance your reactions. Ensure that all members of the family are well-versed in the protocols for different scenarios.
     - **Continuous Learning:** Remain informed on emerging survival methodologies, technologies, and optimal strategies. Revise your plan in accordance with insights gained from exercises and actual events.

This offers a complete strategy for adapting your survival plan to various situations, guaranteeing that you are ready for a range of unforeseen circumstances. By employing these tactics, you can uphold adaptability and efficiently react to evolving circumstances.

Reacting to Novel Dangers and Obstacles

When faced with new risks or problems, it is imperative to respond in a manner that guarantees ongoing safety and adaptability. This section offers techniques for recognizing, assessing, and adjusting to developing risks.

1. **Identifying New Threats**
   - **Monitoring Information Sources:**
     - **News and Alerts:** Stay informed by regularly accessing trustworthy news outlets, receiving emergency notifications, and following local warnings to recognize potential risks as they arise.
     - **Community Reports:** Take heed to community reports and input from neighbours or local organizations regarding new threats or incidents.
   - **Recognizing Warning Signs:**
     - **Behavioural Changes:** Monitor alterations in the surroundings or conduct that could suggest emerging dangers, such as atypical activity or indications of instability.
     - **Environmental Indicators:** Stay vigilant for environmental fluctuations such as alterations in meteorological trends, heightened levels of contamination, or atypical animal conduct that may indicate emerging hazards.

2. Evaluating the Impact

- **Risk Assessment:**

  - **Impact Analysis:** Evaluate the effect of the new threat on your existing readiness and strategy for survival. Take into account the possible impacts on your resources, safety protocols, and overall strategy.

  - **Vulnerability Assessment:** Analyse the weaknesses in your present configuration that could be targeted by the new attack. Identify locations that require greater actions or resources.

- **Prioritizing Response Actions:**

  - **Critical Needs:** Arrange actions in order of importance by considering the urgent requirements and possible consequences of the emerging threat. Direct your attention towards areas that demand immediate action in order to reduce hazards.

  - **Resource Allocation:** Efficiently distribute resources to successfully combat the emerging threat. Revise your resource management plan to reflect fluctuations in availability or demand.

3. Adjusting Your Plan

- **Modifying Procedures:**

  - **Updated Protocols:** Modify your current procedures and policies to specifically target the newly identified threat. Enforce new safety protocols or modify existing ones to enhance their efficacy.

  - o **Emergency Plans:** Revise your emergency plans to incorporate precise strategies for addressing the newly identified threat. Ensure that all family members are informed about the modifications and are knowledgeable on how to respond appropriately.

- **Resource Management:**

  - **Replenishment and Storage:** Revise your approach to resource allocation in order to effectively meet the requirements of emerging needs. Restock inventories and guarantee appropriate storage for new or additional items.

  - **Alternative Resources:** Consider investigating alternative resources or providers in the event that conventional sources are affected by the new threat. Modify your procurement methods to ensure the acquisition of essential products.

4. Enhancing Preparedness

- **Upgrading Equipment:**

  - **New Technologies:** Contemplate integrating innovative technology or equipment that can effectively mitigate the impending hazard. Assess potential strategies to improve your readiness and ability to respond effectively.

- **Training and Skills:** Enhance your training and skills to proficiently tackle emerging difficulties. Explore supplementary resources or enroll in extra courses to enhance your knowledge and skills.

- **Building Resilience:**
  - **Stress Testing:** Perform stress tests on your preparedness strategies to assess their efficacy in mitigating the impact of the new danger. Analyze and identify areas that require improvement, then make necessary adjustments appropriately.
  - **Community Engagement:** Interact with community organizations or local groups to exchange information and cooperate on response endeavours. Enhance collective resilience by strengthening communal bonds.

5. **Communication and Coordination**

   - **Effective Communication:**
     - **Information Sharing:** Engage in communication with family members and pertinent individuals with the newly identified threat and the revised strategies for response. Ensure that all individuals are adequately informed and properly prepared.
     - **Coordination with Authorities:** Collaborate with local authorities or emergency services to synchronize your reaction with official rules and resources.

   - **Maintaining Connectivity:**
     - **Reliable Channels:** Utilize dependable communication channels to remain connected and well-informed. Make sure you have contingency plans for communication in the event of interruptions.
     - **Emergency Contacts:** Revise and uphold a roster of emergency contacts, encompassing local authorities, medical practitioners, and neighbours.

6. **Evaluating and Learning**

   - **Post-Incident Review:**
     - **Assessment:** Once the threat has been mitigated, it is essential to carry out a comprehensive evaluation of your reaction and preparedness endeavours. Assess the successful aspects and pinpoint areas that can be enhanced.
     - **Lessons Learned:** Record the knowledge gained from past experiences and integrate it into your strategies for being ready. Utilize this information to improve future reactions and strengthen resilience.

   - **Continuous Improvement:**
     - **Plan Updates:** Consistently revise your preparedness strategies to incorporate fresh perspectives and evolving circumstances. Remain proactive in adjusting to emerging risks and problems.

- **Training and Drills:** Implement frequent training sessions and exercises to rehearse reactions to emerging dangers. Consistently enhance your abilities and practices to enhance effectiveness.

This section aims to guarantee that you possess the necessary skills and resources to successfully address emerging threats and challenges, so ensuring safety and adaptability in the presence of ever-changing dangers. By applying these tactics, you may adjust to evolving circumstances and improve your readiness for future catastrophes.

Acquiring Knowledge from Actual Bug-In Incidents

Experiences of encountering bugs in real-life situations provide useful insights and lessons that help improve your readiness and techniques for dealing with them. This part examines practical insights derived from real bug-in situations, encompassing case studies and essential lessons for enhancing your own bug-in strategy.

1. **Case Studies of Successful Bug-Ins**
   - **Case Study 1: Urban Blackout**
     - **Scenario:** A metropolitan area endured a prolonged disruption of electricity as a result of an intense storm.
     - **Actions Taken:** Residents who successfully remained in their homes during the crisis had prepared in advance by acquiring alternative power sources, sufficient supplies of food and water, and reliable communication systems. They utilized generators to sustain vital appliances and participated in community support networks.
   - **Lessons Learned:**
     - **Backup Power:** Dependable backup power sources are vital for preserving critical systems during periods of power failure.
     - **Community Support:** Robust community networks can offer vital aid and resources in times of disaster.
   - **Case Study 2: Natural Disaster**
     - **Scenario:** A hamlet located in a region prone to hurricanes experienced a direct impact from a significant cyclone.
     - **Actions Taken:** Efficient bug-in tactics encompassed fortifying residences, accumulating non-perishable sustenance, and establishing secure areas within dwellings. In addition, the residents devised communication strategies and collaborated with nearby emergency agencies.
   - **Lessons Learned:**
     - **Home Reinforcement:** Strengthening your home can offer essential safeguarding against extreme weather conditions.

- **Preparedness Planning:** Thorough strategic planning and effective allocation of resources are crucial for ensuring long-term survival in the face of natural calamities.

- **Case Study 3: Civil Unrest**
    - **Scenario:** Demonstrations and social disorder resulted in heightened security vulnerabilities in an urban region.
    - **Actions Taken:** Effective bug-ins entailed strengthening residences, vigilantly tracking news and social media for updates, and remaining indoors to evade confrontation. In addition, the residents employed monitoring systems to oversee their land.

- **Lessons Learned:**
    - **Home Security:** Heightened security protocols and increased monitoring are crucial in circumstances characterized by civil disturbance.
    - **Information Management:** Obtaining information from several sources aids in making prompt and efficient decisions.

2. Common Challenges Faced During Bug-Ins

- **Challenge 1: Resource Scarcity**
    - **Issue:** Scarce availability of food, water, and medical supplies can put a burden on resources.
    - **Solutions:** Develop and execute tactics to optimize the utilization of resources, create robust supply chains, and explore the possibility of engaging in reciprocal trade with nearby individuals or communities if necessary.

- **Challenge 2: Psychological Stress**
    - **Issue:** Prolonged seclusion and challenging circumstances might have an impact on one's emotional well-being.
    - **Solutions:** Formulate effective coping mechanisms, sustain regular communication with close acquaintances, and participate in endeavours that foster psychological welfare.

- **Challenge 3: Security Threats**
    - **Issue:** The presence of looters or intruders can provide substantial security issues.
    - **Solutions:** Enhance residential security measures, employ advanced monitoring systems, and establish collaborative partnerships with neighbours to ensure collective protection and heightened awareness.

3. Adapting Lessons to Your Plan

- **Incorporating Best Practices:**
  - **Resource Management:** Implement optimal strategies observed in successful instances, such as effective utilization of resources and strategic accumulation methods.
  - **Home Fortification:** Implement practical reinforcing methods based on empirical experiences, such as fortifying doors and windows.

- **Improving Preparedness:**
  - **Training and Drills:** Engage in frequent drills to rehearse and refine reactions to different events. Utilize insights gained from practical experiences to enhance and optimize your protocols.
  - **Community Engagement:** Enhance community connections and develop local networks of support by using effective solutions identified in case studies.

- **Enhancing Flexibility:**
  - **Adaptation:** Ensure that you are ready to modify your plan in response to updated data and changing risks. Utilize knowledge gained from practical experiences to enhance adaptability and durability.
  - **Continuous Learning:** Keep yourself updated on evolving trends and experiences. Utilize continuous learning to revise and improve your bug-fixing procedures.

4. Documenting and Sharing Experiences

- **Personal Documentation:**
  - **Recording Lessons:** Maintain a comprehensive log of your personal bug-in encounters, documenting the difficulties encountered and the strategies employed to overcome them. This material can enhance the effectiveness of your preparation plan and offer useful perspectives for future events.

- **Community Sharing:**
  - **Knowledge Exchange:** Share your personal anecdotes and valuable insights with fellow members of your community. Participate in conversations and forums to share knowledge and tactics for successful bug-in situations.

This section offers pragmatic insights and practical lessons derived from actual bug-in situations, enabling you to improve your readiness plan and bolster your capacity to tackle diverse difficulties. By assimilating knowledge from others and incorporating effective methodologies, you can enhance your ability to withstand and prepare for forthcoming emergencies.

CONCLUSION

Converting your home into a safe and self-sustaining sanctuary during a crisis involves meticulous preparation, commitment, and ongoing enhancement. By applying the tactics delineated in this book, you are actively taking measures to guarantee the security and welfare of both yourself and your cherished ones.

Key Takeaways

1. Preparation is Paramount:

Thorough readiness entails assessing possible risks, accumulating necessary resources, and fortifying your home. The objective is to establish a robust setting in which you can proficiently handle emergencies without depending on outside aid.

2. Adaptability and Flexibility:

Adapting to new dangers and changing conditions is essential. Continuously revise your goals and procedures in response to actual experiences and developing dangers. Flexibility guarantees the ability to effectively address unexpected obstacles.

3. Community and Support Networks:

Establishing robust relationships with neighbours and local communities improves your capacity to effectively address situations. Engaging in collaboration and resource sharing with others enhances the robustness of the support system in times of crises.

4. Mental and Emotional Resilience:

Mental readiness is equally crucial to physical readiness. Create effective techniques to cope with stress, uphold morale, and promote mental well-being in situations of prolonged isolation or uncertainty.

5. Continuous Improvement:

Preparedness is a continuous endeavour. Consistently assess and improve your preparations, carry out practice exercises, and stay updated on advancements in survival methods and technologies. Gaining knowledge from practical encounters and adjusting your strategy can enhance your preparedness.

3. **Adapting Lessons to Your Plan**

   - **Incorporating Best Practices:**

     - **Resource Management:** Implement optimal strategies observed in successful instances, such as effective utilization of resources and strategic accumulation methods.

     - **Home Fortification:** Implement practical reinforcing methods based on empirical experiences, such as fortifying doors and windows.

   - **Improving Preparedness:**

     - **Training and Drills:** Engage in frequent drills to rehearse and refine reactions to different events. Utilize insights gained from practical experiences to enhance and optimize your protocols.

     - **Community Engagement:** Enhance community connections and develop local networks of support by using effective solutions identified in case studies.

   - **Enhancing Flexibility:**

     - **Adaptation:** Ensure that you are ready to modify your plan in response to updated data and changing risks. Utilize knowledge gained from practical experiences to enhance adaptability and durability.

     - **Continuous Learning:** Keep yourself updated on evolving trends and experiences. Utilize continuous learning to revise and improve your bug-fixing procedures.

4. **Documenting and Sharing Experiences**

   - **Personal Documentation:**

     - **Recording Lessons:** Maintain a comprehensive log of your personal bug-in encounters, documenting the difficulties encountered and the strategies employed to overcome them. This material can enhance the effectiveness of your preparation plan and offer useful perspectives for future events.

   - **Community Sharing:**

     - **Knowledge Exchange:** Share your personal anecdotes and valuable insights with fellow members of your community. Participate in conversations and forums to share knowledge and tactics for successful bug-in situations.

This section offers pragmatic insights and practical lessons derived from actual bug-in situations, enabling you to improve your readiness plan and bolster your capacity to tackle diverse difficulties. By assimilating knowledge from others and incorporating effective methodologies, you can enhance your ability to withstand and prepare for forthcoming emergencies.

CONCLUSION

Converting your home into a safe and self-sustaining sanctuary during a crisis involves meticulous preparation, commitment, and ongoing enhancement. By applying the tactics delineated in this book, you are actively taking measures to guarantee the security and welfare of both yourself and your cherished ones.

Key Takeaways

1. Preparation is Paramount:

Thorough readiness entails assessing possible risks, accumulating necessary resources, and fortifying your home. The objective is to establish a robust setting in which you can proficiently handle emergencies without depending on outside aid.

2. Adaptability and Flexibility:

Adapting to new dangers and changing conditions is essential. Continuously revise your goals and procedures in response to actual experiences and developing dangers. Flexibility guarantees the ability to effectively address unexpected obstacles.

3. Community and Support Networks:

Establishing robust relationships with neighbours and local communities improves your capacity to effectively address situations. Engaging in collaboration and resource sharing with others enhances the robustness of the support system in times of crises.

4. Mental and Emotional Resilience:

Mental readiness is equally crucial to physical readiness. Create effective techniques to cope with stress, uphold morale, and promote mental well-being in situations of prolonged isolation or uncertainty.

5. Continuous Improvement:

Preparedness is a continuous endeavour. Consistently assess and improve your preparations, carry out practice exercises, and stay updated on advancements in survival methods and technologies. Gaining knowledge from practical encounters and adjusting your strategy can enhance your preparedness.

Final Thoughts

When you begin the process of turning your house into a safe and protected place, keep in mind that being prepared, adaptable, and resilient are crucial for effectively managing any situation. The information and tactics presented in this book establish a solid framework for constructing a resilient bug-in plan capable of withstanding diverse circumstances.

Your dedication to being prepared demonstrates a proactive and responsible attitude towards safeguarding the well-being and protection of your home. By implementing the knowledge acquired and consistently enhancing your preparedness, you are making a substantial investment in your future welfare.

Remain watchful, stay well-informed, and ensure you are ready for any situation. The sense of tranquillity that accompanies the knowledge of being fully prepared for any obstacle is immeasurable. By implementing these principles, you are not only establishing a secure environment for yourself, but also enhancing the robustness and adaptability of your community.

We appreciate your decision to undertake this journey towards enhanced preparation. May your endeavours result in a future that is more safe and resilient for both yourself and the people you hold dear.

Made in the USA
Monee, IL
02 September 2024